True Round
Metal Boat Building

via

Approximate Development

DL Schaffer

The findings detailed in this book reflect the authors' own opinions and research. The author assumes takes no responsibility for the use or misuse of the information contained herein.

Introduction:

The steel and aluminum True Round boatbuilding method that I am about to introduce, **True Round Metal Boat Building - By way of Approximate Development**, does not entail the use of an English Wheel, Line Heating, or any other vague or artistic like metal forming techniques.

As you shall see, **True Round Metal Boat Building - By way of Approximate Development**, pre-engineer's the entire metal boat structure, including the Transverse frames, Longintudinal frames, and shell plating.

For those who may ever heard of **Approximate Development**, it is the method used every day in the Metal fabrication industry used to Unfold three-dimensional sheet-metal products onto a single plane for fabrication by bending and rolling methods without stretching or shrinking the metal.

Since a True Round metal hull design is also a three-dimensional sheet-metal product it also an be developed onto a single plane.

With the unfolded patterns in hand a Builder can fabricate and True Round metal hull as simply as *a' Childs Erector Set*.

Approximate Development is simple, consistent, and predictable. Gone are the artistic and obscure method of the past.

About Myself:

The process of combining computer technology with 'Approximate Development' to metal boat building is a result of my career in Architectural metal fabrication, Cnc Programming, Sheetmetal pattern development, and my studies at the 'Westlawn Institute of Marine Technology.

Although I have built hard chined steel designs previous to building the prototype build shown below. I had no experience building a true round metal hull. I have no arcane or artistic like skills.

What I do have is a complete understanding of the theory of 'Approximate' Sheetmetal pattern development'.

The build used to illustrate Approximate Development is a small, curvy, classic tumblehome design. If this curvy aluminum fabricated boat can be built using Approximate Development, then any other True Round hull form would certainly be a breeze.

Figure4-5

This book takes the arcane and artistic skills out of plating a True Round metal hull, replacing those skills with time honored and tested sheet-metal fabrication methods use every day in the Metal Fabrication Industry!

DL Schaffer

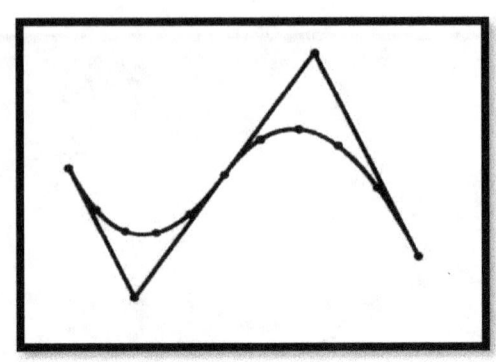

Notes

Materials
Tools
Techniques

Material Grade Selection - Aluminum

Aluminum is classified by alloy series. Only the copper free marine alloys of the 5000 and 6000 series are of concern to the boat builder. The 5000 series is magnesium based while the 6000 series is magnesium and silicon based. Both series have very good weldability and can be welded using both the mig and tig welding techniques. The 5000 series has more malleability making it easier to form to different shapes. Use the 6000 series only when structural shapes, extrusions, and piping are unobtainable in the 5000 series.

The two designations within the 5000 series which are most suitable for yachts and small sailing craft are 5052-H32 and 5086-H32. Of these two alloy designations 5052-H32 is the easiest to work. On the other hand, 5086-H32 is the better grade to use, it is more corrosion resistant, ultimately stronger, a little harder to work with than 5052-H32 and is of course, more expensive.

A high-end aluminum boat would use 5086-H32 through its construction with 6063-T6 used where pipe is required. 5052-H32 would be used only to solve metal forming problems.

To keep material cost down, the medium choice of materials would be 5086-H32 for the hull plating with 5052-H32 used for the deck and house plating. The 5052-H32 would also be used for hull framing. Lastly, 6063-T6 for pipes.

The low-end material choice is to use 5052-H32 throughout construction with 6063-T6 used where pipe is required. In this instance the hull should be painted. When welding 5086-H32, 5052-H32, and 6063-T6 to each other or to themselves the only filler wire alloy to use is 5356.

Material Grade Selection – Steel

Most steel hulls are built using ASTM A36 steel. Commonly referred to as mild steel, hot-rolled steel, plain-carbon steel, and low-carbon steel. Low carbon steel contains approximately 0.05 to 0.15% carbon making it malleable, ductile, and easy to form. It has excellent welding properties and is suitable for grinding, punching, taping, drilling, and machining processes. It is priced relatively low while providing material properties that are more than acceptable for steel boatbuilding. ASTM A36 steel is readily available everywhere in a large variety of shapes: Steel/Plate, Rectangle & Square Bar, Circular Rod,

Channels, Tee Bar, Angle, Beam Shapes and Pipe.

You could forgo sandblasting if you use Pre-primed hot rolled steels, which have been wheel abraded. In theory you only need to touchup the weld seams and other areas where the paint was compromised. Making this a necessary practice at the end of each day's work.

Corten steels have no advantage in steel boatbuilding. They require sandblasting and painting like any other steel use in hull construction.

Cold rolled steel is essentially hot rolled steel that has been processed further to produce a steel that has closer dimensional tolerances, straightness, and a wide range of finished surfaces. Fabrication of custom hatches and window frames would be an application for choosing Cold Rolled Steel.

Sandblasting Steel

Sandblasting is required if using unprotected steel. Inherent to hot roll steel is the mill scale attached to its surface. This mill scale is very hard and difficult to remove by mechanical means.

The best way to deal with this firmly attached scale is to let the hull weather. It takes about six months outside for the weather to remove the mill scale. It is good practice to allow this to happen before sandblasting the boat. Attached mill scale takes twice as long to blast down to bare metal compared to the same area where the hull was allowed to weather.

Tools and Equipment

Steel - Folks that are planning to build a boat probably already have some type of workshop geared to their specialty, be it wood, metal or composites. If so, your workshop may already have many of the tools required to build a Bézier aluminum or steel design.

From my personal experience, with steel builds, all you really need is a straight AC arc welder, an oxygen-acetylene cutting torch or a plasma cutter, five inch and seven-inch hand grinders, Cut-Off Saw, Drill Press, and a variety of clamps and wedges.

Additionally, Electrode Holders used for Stick welding are small in comparison to Mig guns. The ease that you can get into tight, distant, and deep spaces by simply bending the welding rod to another angle is priceless. 6011 is an all-purpose welding rod used for general repair and fabrication. 6011 welding electrodes penetrates deep into the base metal.

Aluminum requires more sophisticated welding equipment. A mig machine with spool gun for heavy hull work and a tig machine for lighter work. For aluminum construction a vertical Band saw can replace the plasma cutter.

Fabrication Techniques:

Full size patterns and Cnc cutting files add versatility to Bezier Chine Designs by offering a choice of several cutting methods in the pre-fabrication phase of the build that can save time and reduce cost. Such considerations are:

Cnc Cutting vs Hand Cutting - Some Builders preferred Cnc cutting for both steel or aluminum parts. It is a time saver when your budget allows the additional expense. Other builders take a more traditional approach and prefer to manually cut their steel or aluminum parts.

Manual Saw Cutting - is used mostly in aluminum construction since the metal cuts easily.

Manual Plasma Cutting - is used mostly to cut steel, however aluminum can be Plasma cut.

Manual Roll Forming - works equally well for steel or aluminum. It is an efficient way to form Longitudinals, deck beams and sections of transverse frames.

Full size patterns and Cnc cutting files are interchangeable formats to accommodate the builder's working preferences.

For example, the Bezier 12.5, an aluminum build, had no parts cut using Cnc equipment. The shell plating, center vertical keel, sheer and deck Longitudinals are saw cut, while the hull Longitudinals between the sheer line and Cvk were rolled formed.

Working with Patterns

When transferring patterns be sure to smooth out any wrinkles. Patterns are firmly held in place with weights. Mark around the edges with a wide marker keeping the marker in contact with both the pattern and the material.

A small part shown below, has been marked by the above method. It is being saw cut within the 'marked' lines, removing the line in the process. By removing the 'marked' lines you are cutting at the edge of the template.

In the part, shown below, notice the area of the 'Black' still remaining along the top edge of this part. Any part of the 'marked' line remaining will be a high spot, and needs to be removed.

The opposite extreme, bottom edge, shows undercutting where too much material has been removed. Undercutting is to be avoided; you cannot easily add material!

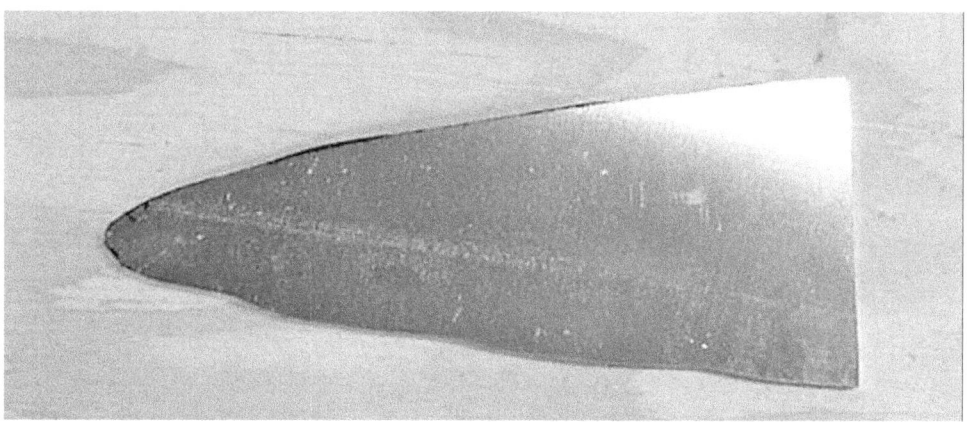

Squaring Techniques

Corner to corner or diagonal measurements are used to assemble and make all assemblies symmetrically square. The illustration, Below, shows two rectangular frame shapes, both measure 20" x 10". The top frame has been squared. All the corner angles are 90 degrees, therefore, its diagonal dimensions corner to corner are equal at 22.360".

The bottom frame is out of square by 2.5 degrees, its angular corner dimensions are other than 90 degrees. The diagonal dimensions are 22.747" for the long side and 21.967" for the short side.

To bring the frame back to the desired symmetrical - rectangular shape, subtract the smaller dimension from the larger, then divide by two. This result would either be added to the smaller dimension or subtracted from the larger dimension.

If this were a real rectangular frame, we would move the sides to a position where both diagonal measurements were equal. The result would be a shape whose corners were all 90 degrees while maintaining the perimeter dimensions of the rectangle.

This basic technique can be applied to any other complex irregular shape, such as a transverse hull frame, as it does for this simple rectangle.

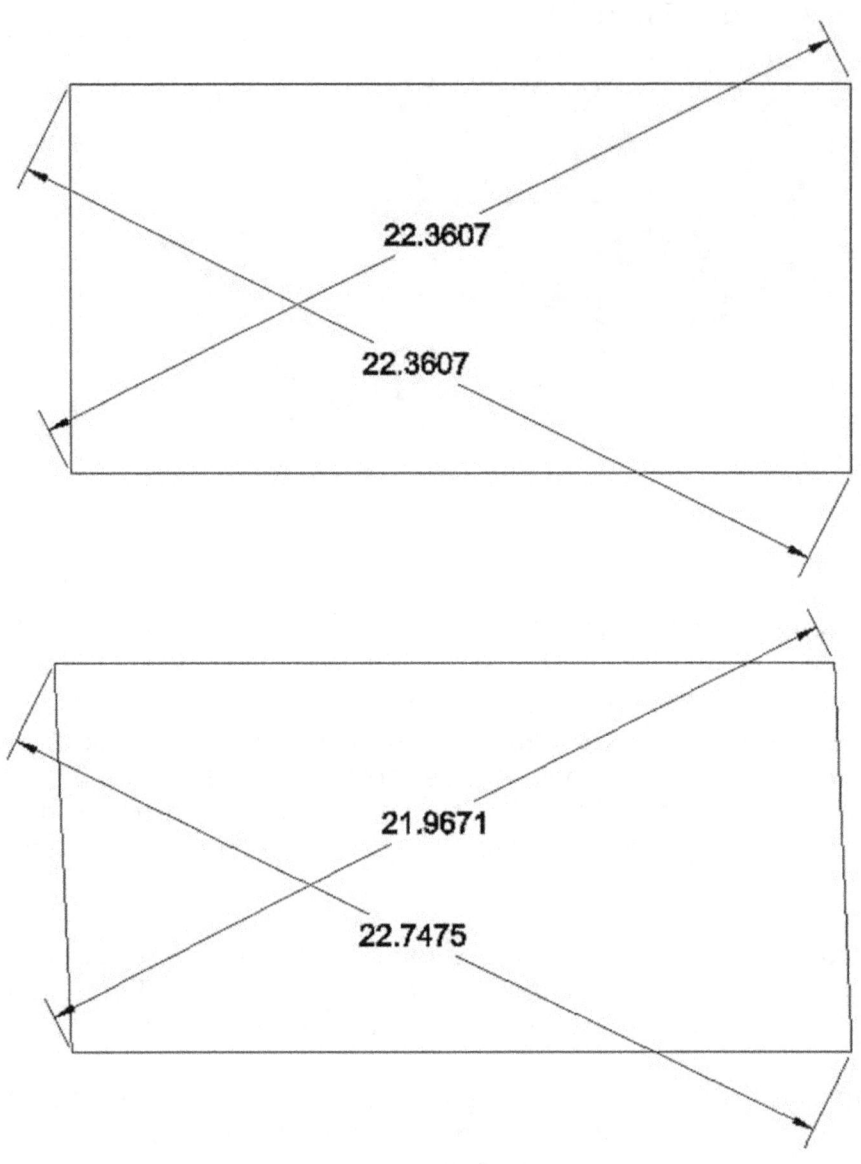

Notes

Drawings
And
Patterns

What in the Plans:

Plans are provided to the Builder in Three Forms:

- Architectural Drawings – PDF format
- Full-Size Patterns – PDF format
- Cnc files – DXF format

Architectural Drawings:

A sampling of Architectural files for the Bezier 12.5 are show below.

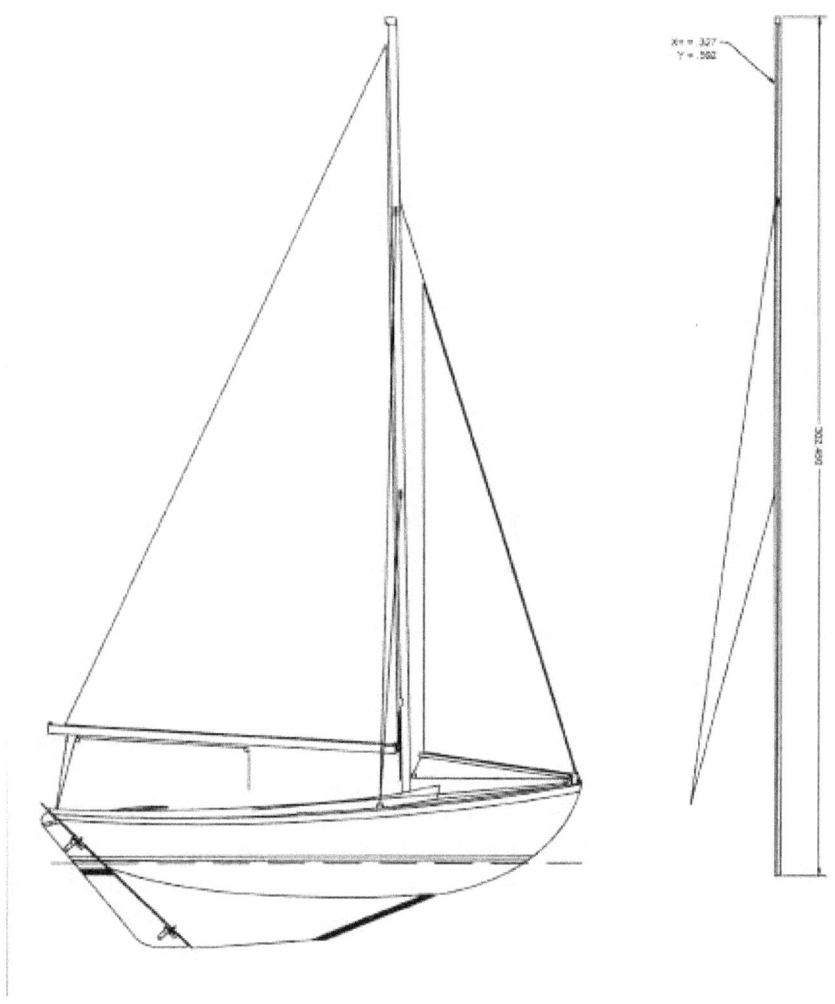

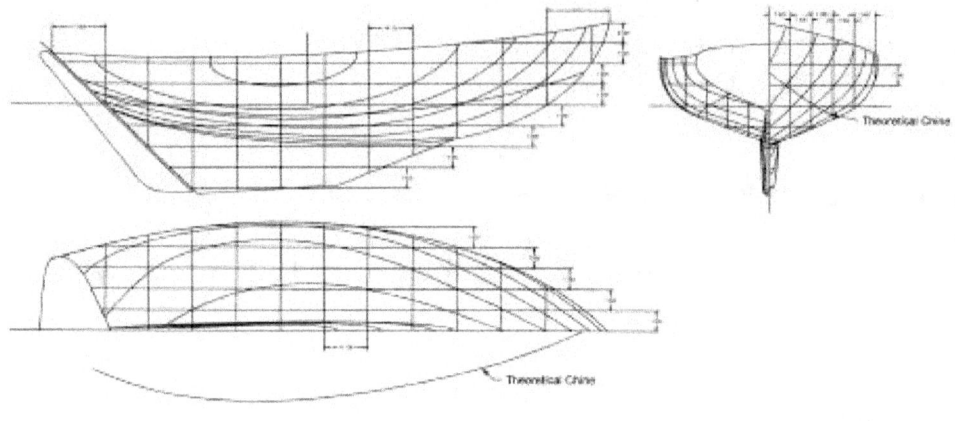

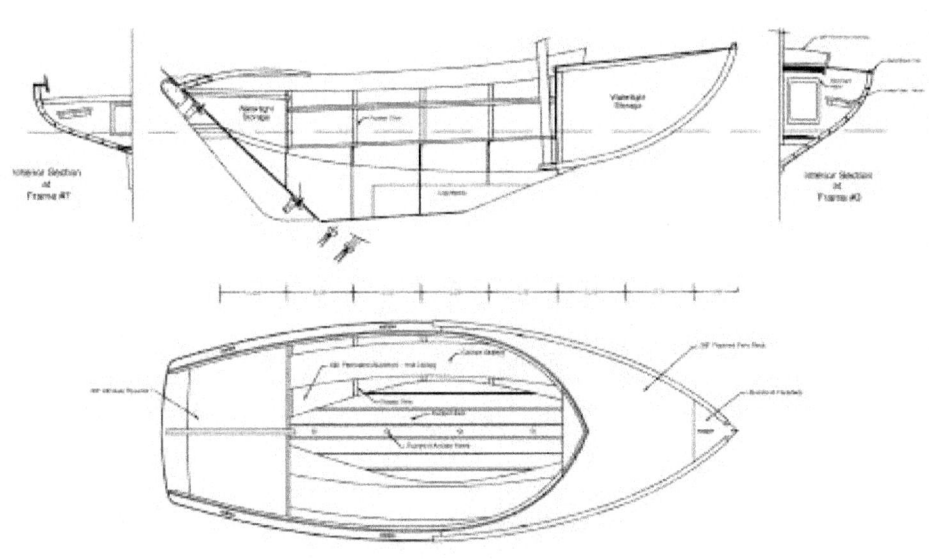

Full-size Patterns:

Full size patterns are provided in PDF file format and contain essential fabrication information that cannot be conveyed by Cnc cutting files.

Simply take the PDF's to an office supply store with a large format printer. Scaled architectural drawing will be printed to the scale given on the drawing, while full-size pattern will be printed full-size.

After printing store at room temperature and keep dry. Do not store in a damp environment. Always take care to prevent folds and creases.

The illustration below is a representation of ten (10) full-size shell plate sections plotted onto a 36" x 96" sheet. Note that an independent 'Check Dimension' line has been run diagonally from a point near the upper left corner to a point near the lower right corner. It is dimensioned at 96.000".

Since the plot is full-size, measure this 'Check Dimension' line. It must be 96.000" by your tape measure. Now be sure the plot is on a flat surface free of wrinkles.

If a 'Check Dimension' is other than 96.000" there may be an error in the calibration of the Large Format plotter.

For example, let say there is an error of + or − 0.125" over the 96.000". This does not mean that every shell plate is 0.125" larger or smaller than it should be.

Let's say that the shell plates are 8.000" wide, therefore 12 times shorter than 96.000". If we divide 0.125" by 12, the result is 0.010", definitely within working tolerance.

Let's say the error was 0.500" over 96.000". the error of the say shell plate would be 0.500/12 = 0.041" or a little more than 1/32"

What I am saying is to check the actually plot for accuracy before proceeding. See the below illustration.

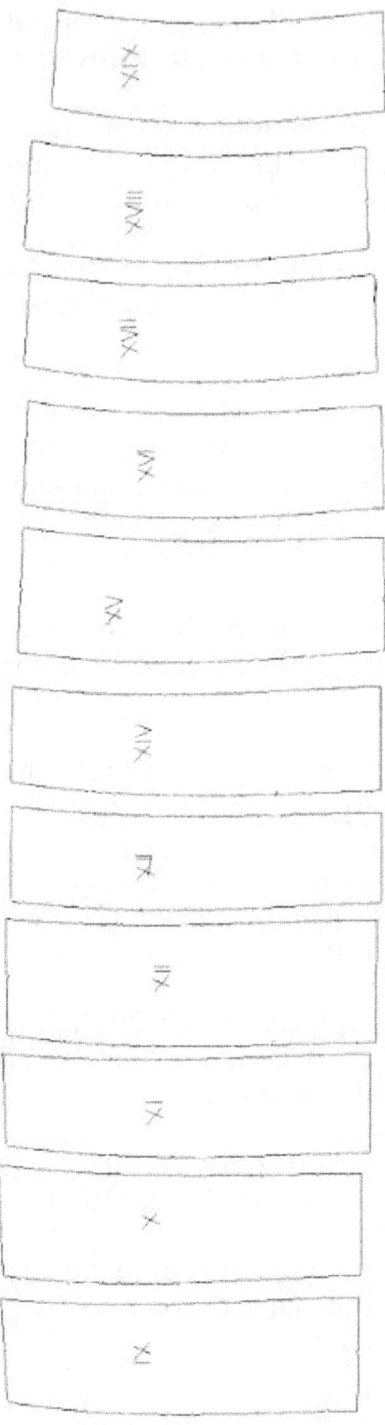

Cnc cutting files:

There are three types of Cnc cutting machines: Plasma, laser and water jets. Plasma machines are probably the most common but Laser and water jets can hold a closer tolerance than plasma machines. Lasers cut faster than water jets. Water jets can cut just about any material.

These machines are 'Vector' orientated. They do not have calibration concerns of Large Format printer's which are 'Raster orientated.

- You are free to contract with any fabrication shop, and negotiate the cutting price. Cnc cutting rates are priced by the quality (tolerance) of the cut and length in inches.

- Check with your cutting shop on the tolerance that their cutting equipment will hold. A tolerance of plus or minus .015" would be acceptable. Closer tolerance are possible, however with an increase in machine time therefore increasing cost.

- Another consideration is the size of the CNC machines table. A four foot by ten-foot table is absolute minimum. The Cnc cutting files provided are easily manipulated by all machine software programs for any optimization or nesting requirements particular to its table size that save machine time and material.

- The cost of paper when plotting full-size pattern using a large format plotter is non-existent as compared to the cost of steel and aluminum sheet, therefore the various parts need to be 'Nested' into close proximity of each other. Your Metal Fabricator has the software to limit the waste when cutting from sheet material.

Notes

Pre-Fabricating
the
Framing

Bezier Chine design and Construction is exact. All parts of the hull can be Pre-Fabricated before actual construction of the hull begins.

The parts can be CNC cut, hand Saw Cut, or Roll Formed using either the Cnc cutting files or the Full-size patterns.

Transverse Framing

An Architectural drawing of each Transverse frame similar to the illustration below provides all the dimensional information necessary to assemble each frame.

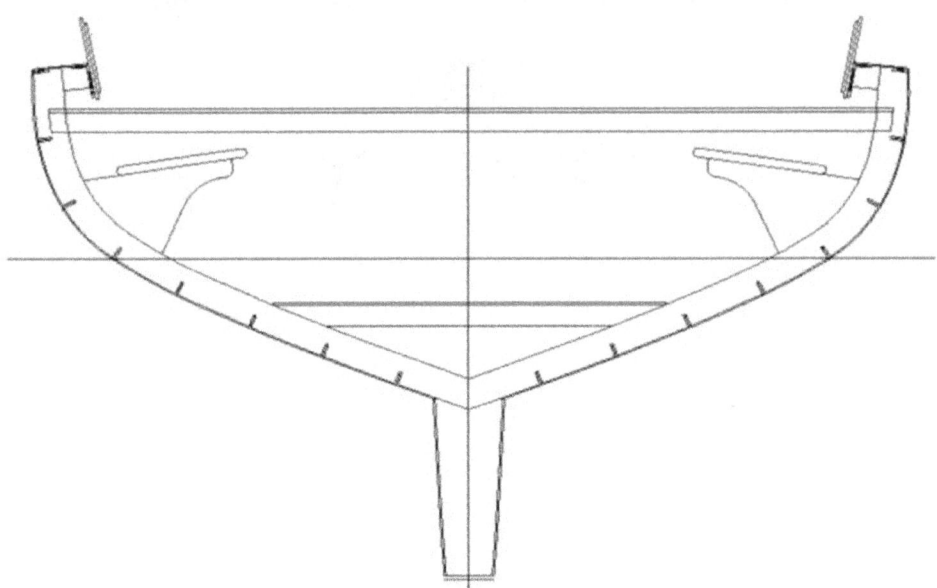

During assembly the parts of the Transverse frames are held loosely together by small tack welds, as you employ the squaring techniques previously describes.

If any frame does not true-up up dimensionally at any location, look for the reason. Find the part which is too long or short, a notch cut too deep or not deep enough, or a high or low spot. Time taken here will be well rewarded later.

Longitudinal Framing

Roll forming the longitudinal frames from Flat Bar significantly lowers material cost. There is no waste.

Longitudinal frames span the extent of the hull, making them difficult to work in one piece. It is reasonable to divide the Longitudinal's into practical working lengths.

Metal construction allows the Longitudinals to be easily spliced back together during hull fabrication. The splice should be a minimum of four inches from any transverse frame, and a little extra length should be added to allow the spliced Longitudinal a place to overlay each other during the fabrication purposes.

Longitudinal Rolling Fixtures

A light duty two-point rolling device, shown below, was fabricated to roll-form the 1" x ¼" aluminum flat bar Longitudinals frames of the Bezier 12.5.

It consists of two 1/2" pins welded to a base plate a short distance apart. End plates are welded to the pins to keep the flat bar from twisting out during the forming process. The device is shown attached to a stable platform.

Before rolling your first part be aware that the actual working cut size will be the length of the finished part plus the distance between the pins at each end. This particular roll forming device requires a minimum of eight inches of additional material, since the distance between pins is four inches.

To use, insert the flat bar between the two pins. Downward hand pressure applied to one side of the flat bar will pivot the bar and engage the second pin. With enough downward hand-pressure the bar will kink.

When a suitable kink is achieved, release the pressure. Move and repeat the process again and again over a suitable length, estimating the curve result. Smooth curves are achieved by small kinks placed close together.

Next, place the flat bar on the template, shown below, to check the alignment between the work and the full-size pattern. If the curve

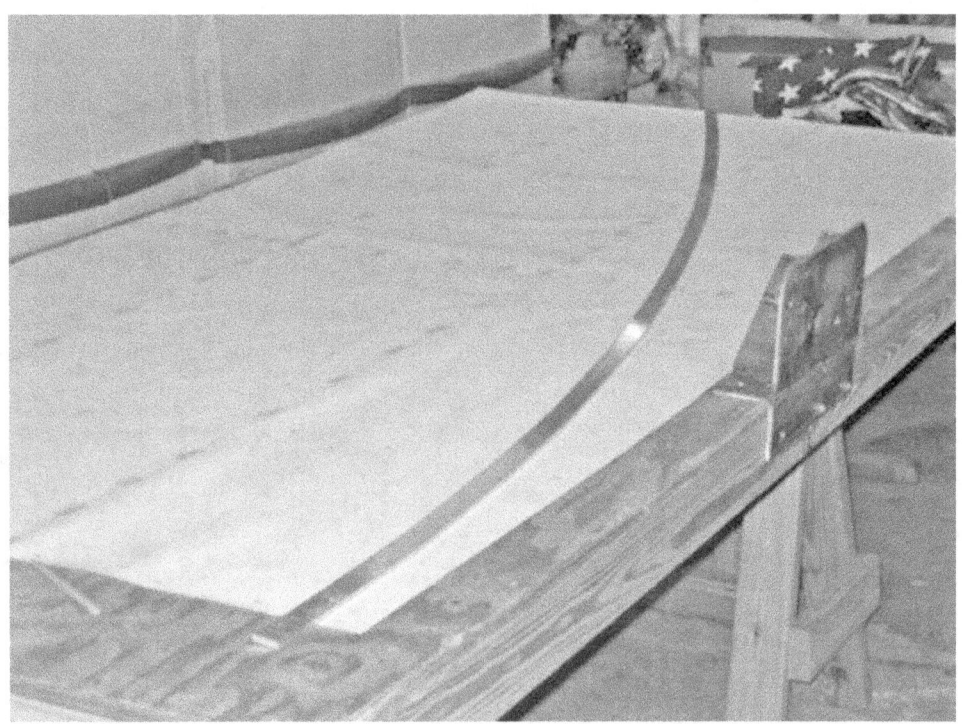

matches the full-size pattern, continue on. If you need more curve, place the flat bar back in the rolling device and repeat the process. If you over-rolled the part turn the flat bar over to remove some of the curve. Roll and check the longitudinal as many times as necessary to obtain the correct curve. Time taken here will be well spent!

For heavy work use the three-point rolling device, shown below, will easily roll 1/2" x 4" steel flat bar. The principle and operation of this three-point device is the same as the light duty rolling machine except a hydraulic jack is used instead of hand pressure alone. Remember to add half the distance between the pins at each end of the part to be rolled.

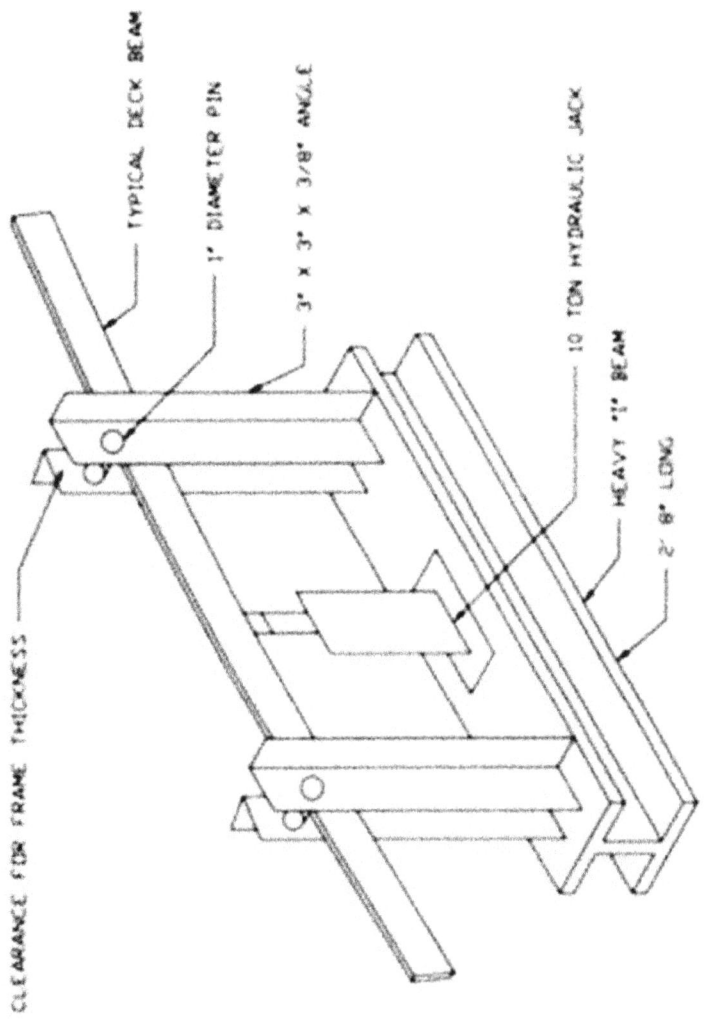

Notes

Pre-Fabricating
the
Shell Plating

Shell Plating – Cnc Files

The shell plating segments are pre-engineered to align with a given location on the hulls framework. They can be Cnc cut or manually cut using the full-size patterns.

The shell plate, shown below, has just been cut by a Cnc water jet.

Shell Plating – Full Size Patterns

A typical shell plate, next page, is shown plotted full size contains the following information:

- The terms 'Sheer' and 'Chine' orientate the pattern.

- The single small half circles along this short side of the pattern align the shell plating to its transverse frame at the 'Sheer' and 'Chine' Longitudinals, if applicable.

- The curved vertical line, near the center of the pattern, locates the intersection of the shell plating and it transverse frame between the aforementioned half circles. It is interesting to note that the curved line will become straight after the pattern has been formed.

- In this particular pattern there are twenty-three (23) 'Element' lines. These lines more importantly indicate where the tooling of the 'Press-break' will engage the material to form the pattern back to its three-dimensional shape. The bend angle chosen for all shell patterns, in this design, is three (3) degrees.

- The small half circles along the vertical edges of the pattern indicate the locations where the shell plating crosses a longitudinal frame.

- The cross lines associated with the numbers 21.937 inches and 1.477 inches are reference dimensions that aid the forming process. They provide the corner to corner finished dimension **after** the shell plating segment has been formed.

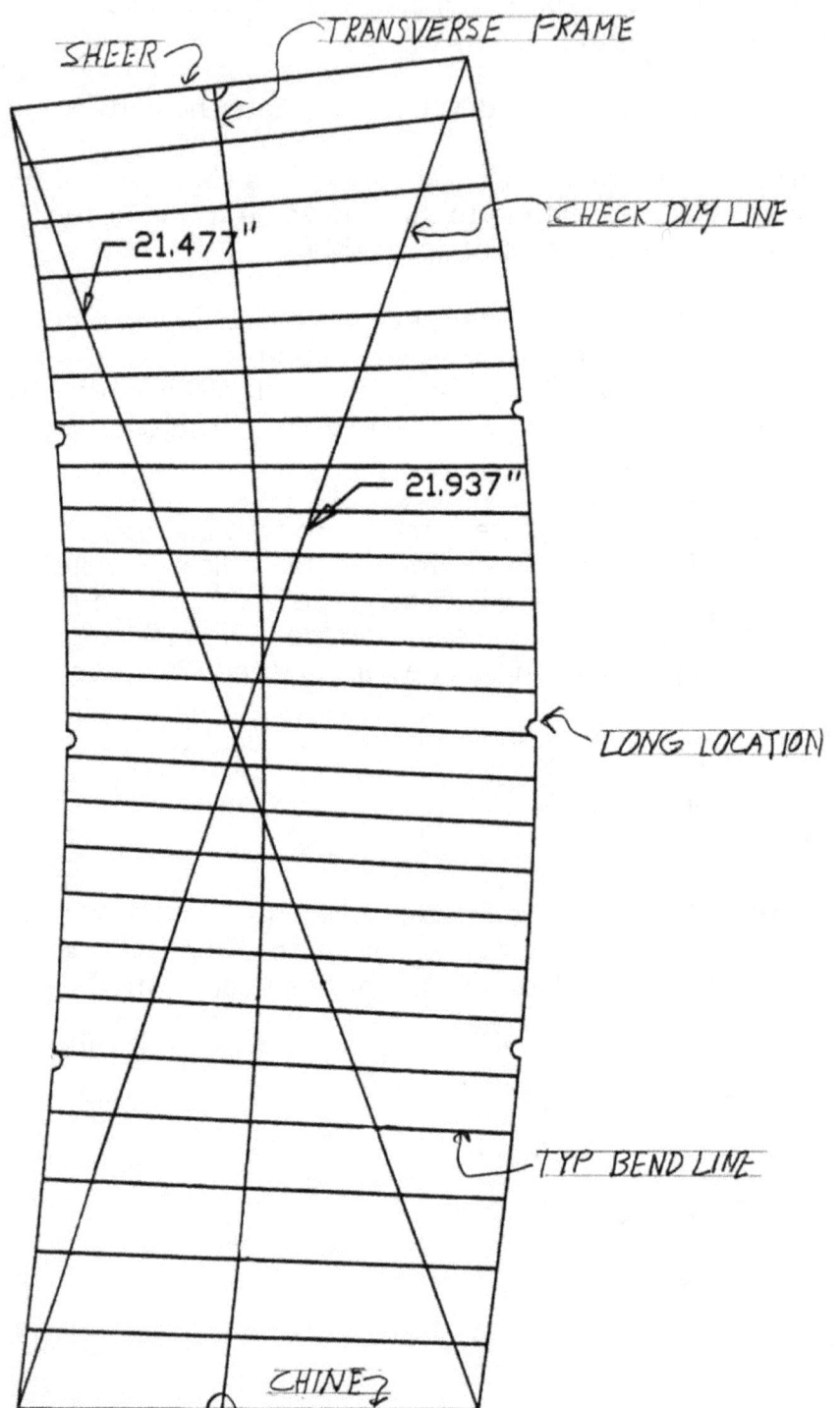

SHEER

TRANSVERSE FRAME

CHECK DIM LINE

21.477"

21.937"

LONG LOCATION

TYP BEND LINE

CHINE

Press-Brakes & Tooling

Bezier plating sections are free formed surfaces and cannot be formed in a rolling machine like a 'Radius Chine' design. Free formed surfaces need to be formed in a 'Press-break'. As with Rolling Machines, every metal fabrication shop has a Press-brake.

'Press-brakes' are very open-ended machines and can accommodate a large variety of tooling. The small 'Press-break' shown below would be more than adequate to form the aluminum or steel shell plating for any 'Bezier design.

To form the shell plating, illustrated on the next page, a round top punch with a vee shaped bottom die works well. The bottom die is fixed to the bed of the press-brake, while the punch is attached to the upper movable ram. For most purposes, the female opening is between one half inch and one inch, while the male punch would best have a radius or rounded tip.

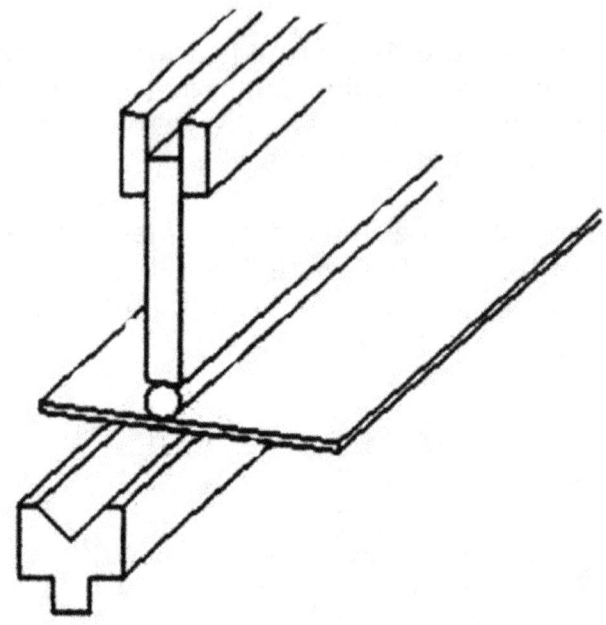

Choosing the correct female die opening is determined by the distance between bends. The material, illustrated below, is ideally supported when no previous bend line is located within the bottom die opening.

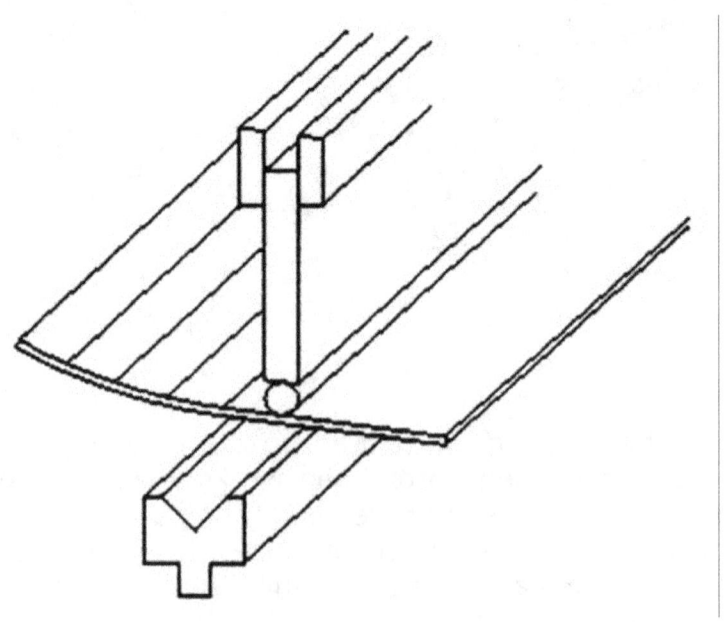

The Illustration below shows a previous bend line within the die opening. This indicates that the die opening is too large. Choose a female die opening that will accommodate the correct condition around 80 percent of the time.

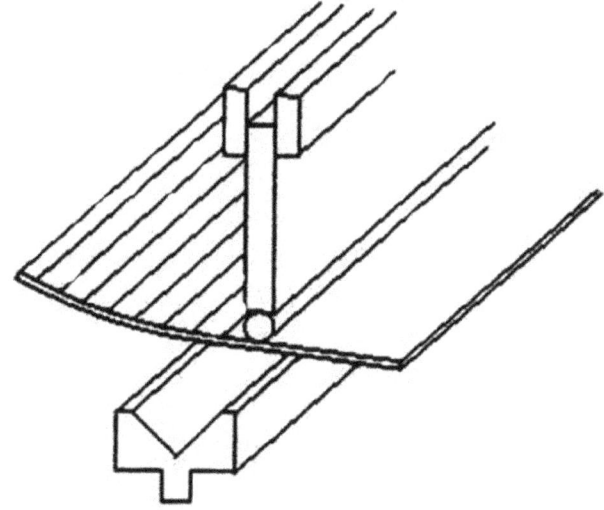

Setting up the Press-brake

An advantage of Bezier Curve Design is that only one bend angle and one press-brake setup is required. Once the bend angle has been set, all the shell sections are formed without readjusting the ram (height setting) of the press-brake.

The quality of the bend angle is critical and is directly related to the condition of the press break, tooling, the fineness of its ram adjustment, and the ability of the machine to hold the bend angle over multiple bending cycles.

Setting up the press break for an accurate bend angle is very important. Errors in the initial setup as small as a quarter degree, will cause the finished piece to be over or under formed. For example: If the initial setup bend angle was over or under by 1/4 degree and a pattern has twenty bend lines the finished section of shell plating would be over or under rolled by 5 degrees; 1/4 degrees x 20 bends = 5 degrees. Do not be intimidated by the tolerances; there is a straight forward four step procedure used to set the ram height.

Step One: Use the same type of material, thickness, and approximate length of the parts to be bent for setting up a 'Test Bend', No exceptions.

Step Two: Estimate the ram height adjustment setting on the press-break and make a test bend, then measure the resulting angle. If the bend angle is over or under, adjust the ram setting up or down. Use a new section of the test material each time another test bend is made.

Step Three: When you are satisfied with the result of a single test bend, further refine the press-brake setup, shown below, by three or more successive bends in a row. Take an angle measurement of the combined bends. If three bends were used the angle will be nine (9) degrees. Further adjust if necessary.

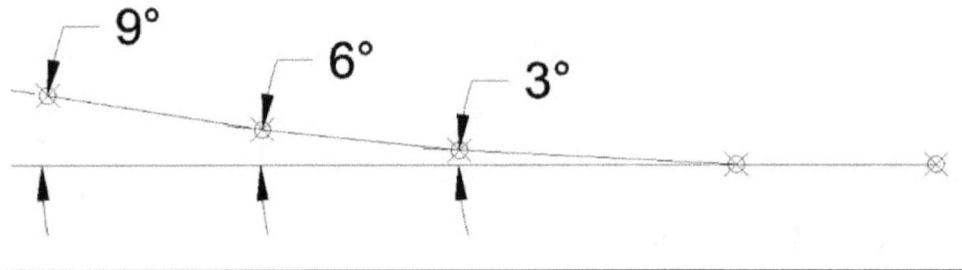

Step Four: The accuracy of the press-brake over numerous bends will be confirmed by using the geometry of a seventy (70) degree section for a ten-inch radius cylinder.

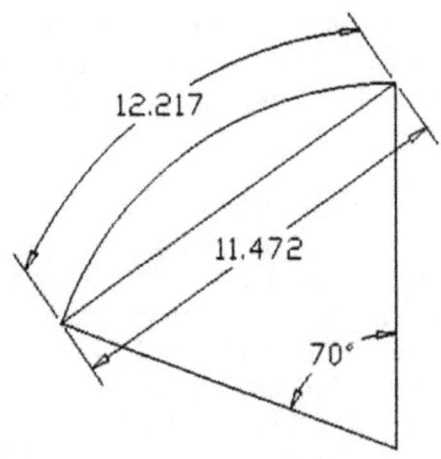

Cut a test piece 12.217" *(Below)* by the approximate width of the average shell section.

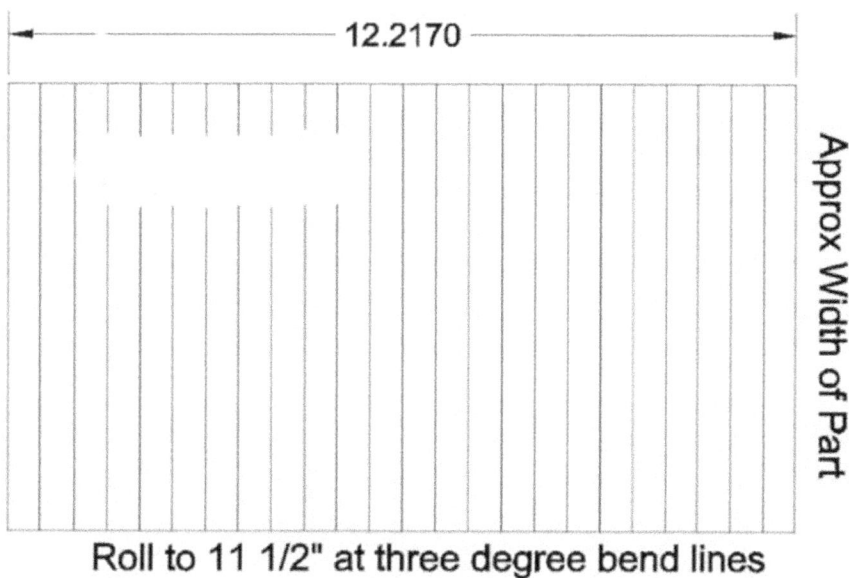

Roll to 11 1/2" at three degree bend lines

Tape the paper patterns directly onto the material being formed. Bend all the 'Elements' in a single run.

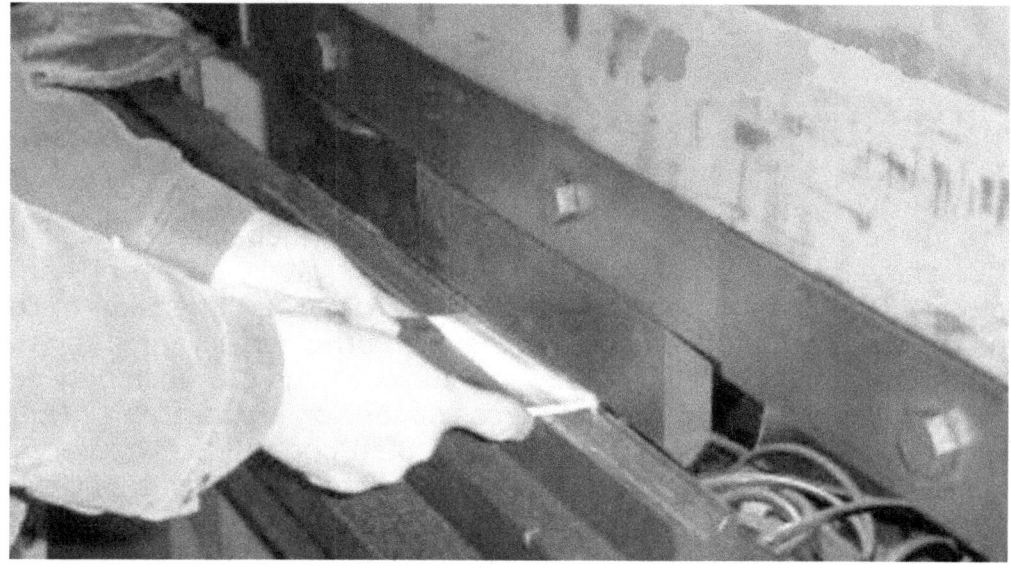

When completed verify the accuracy of the press-break setup by measuring the chord. If the ram setting is correct, the length of the chord should be 11 1/2". If you are over or under this dimension the depth of the press break needs to be adjusted. Adjust until 11 1/2" is reached. ***The press-brake is now set.***

Forming a Shell Plating

With the press-brake set, tape the paper shell plate pattern directly onto the material to pick up the bend lines. Run the whole section thru hitting all the bend lines in sequence only once. Do not double hit any of the bend lines. Start forming the plate from either end towards the middle. If the work becomes difficult to handle turn it end for end, working again toward the middle.

With the plating section formed, it can be verified for accuracy by referencing to the cross-checking dimensions given for every paper pattern. The cross dimension noted in are ***not*** the dimension of the pattern on the flat, they are dimensions corner to corner after the pattern has been formed. An error of plus or minus 1/8" inch between the formed part and the indicated cross dimension is acceptable. Note, that is always better to over form a part than under forming a part. It is easier to take bend out of an over formed part than it is to put bend in an under formed part.

A plywood template, shown below, scribed from its transverse rame Pattern will also verify the accuracy of the press-brake setup.

With the first plate formed, all the others can follow using this press-brake setting. Check each plate using the cross-reference dimensions and templates where applicable.

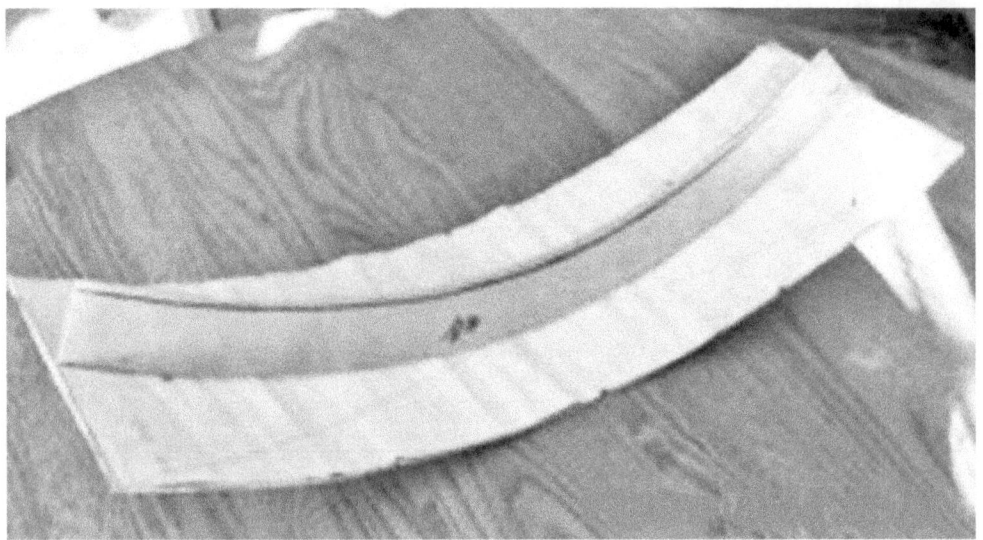

Notes

Keel
Fabrication

The keel is hollow, the shell is supported by the hulls transverse frames. Constructing the keel and hull independent of each other and upside down is the procedure illustrated. The prototype hull used this construction procedure. I now however, recommend that the keel be constructed integral to the hull.

The Full-size pattern for the keel shell plating includes all the reference lines necessary to form and located the keel onto the keels framework.

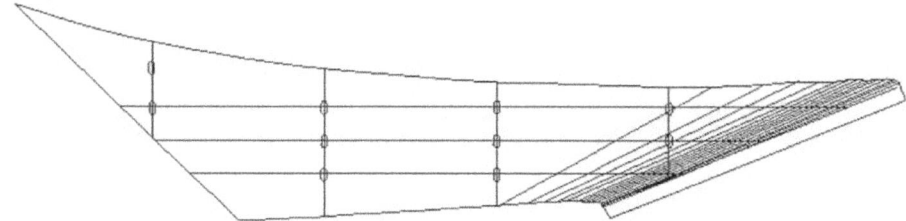

- **Transverse Frame Lines:** Are the vertical lines that align the shell plating with the transverse frames. Obround slots (1.500" x 0.312") are located, on these lines.
- **Plug Welding Obround Slots:** The inside of the keel is generally inaccessible to welding equipment, therefore 'Plug' welds are used to secure shell plating to the transverse frames from the outside.
- **Bend Lines:** The angled lines on the pattern are three (3) degree bend lines. They are used to form the nose section of the keel in a press-break similar to the process used to form the shell plating. Note that additional material has been added to the nose section for bending purposes and will be removed before applying the keels shell plating to the framework.

- **Forming Template Lines:** When aligning the templates **(Figure L)** with the 'Forming Template Lines' gives the Break Operator a good visual on the accuracy of the forming process.

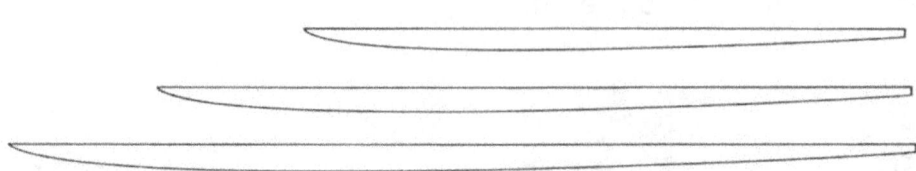

Forming the Keel Plating:

While the same principles of fabrication apply to both keel and hull shell plating, the leading edge of NACA foils naturally have very tight curvatures. The bend lines are so close together that most will fall within the bottom die opening. As previously discussed, this can lead to a breakdown in the accuracy of the forming process.

With this in mind forming the leading edge of the keel shell plating requires some creative press-brake operations, such as sacrificing material by cutting and test bending a full-size mockup of the forward section *(Below)* of the keel, before committing to the much larger full keel section.

It is always good practice to test bend complex parts even if you need to forgo material to facilitate an accurate outcome in the finished part.

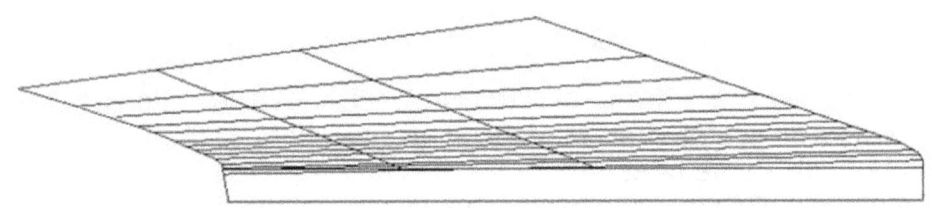

Notes

Setting up
The
Framework

Setting up the Framework

'Bezier Chine Design and Construction' designs use a longitudinal framing system. In this system widely spaced Transverse framing support closely spaced Longitudinal frames which in turn support the shell plating, therefore the shell plating is welded only to the Longitudinal frames. Never the Transverse frames.

- Align, Level, Center and Square the transverse frames Two thru Seven onto the building strongback as accurately as possible.
- To these transverse frames add the CVK and the second longitudinal from the Sheer Longitudinal.
- The bow and stern views of the structure, following, are far from stable at this point of fabrication.

At the Bow. Shown below, the second longitudinal is not tied to either the CVK or Transverse frame #1.

At the stern, shown below, Frame #8 and the Transom are not yet installed. The second longitudinal is only secured as far as frame #7 and extends a distance past the transom. The triangle section at the end of the CVK is used to align the transom. It will be removed at an appropriate time.

- Start adding the other longitudinal between Transverse frame Two thru Seven.

- Recheck all locations and intersection of the strongback and the initial framework between transverse frames two thru seven.

- The two Longitudinals nearest the Center Vertical Keel are installed after the pre-fabricated keel is positioned.

- The framing structure should now be somewhat stable between transverse frames two thru seven.

Since the Longitudinals are formed in two sections you can work the bow section separate from the stern section. Allow the bow and stern Longitudinals to overlap near the center of the hull. Splices in longitudinal frames must be at least four (4) inches from any transverse frame.

- At the Bow, shown below, position Transverse Frame One and tie in all the Longitudinals to the CVK.

- At the stern, next page, position frame #8.

- Align the transom to the CVK and the second longitudinal.
  When the transom, CVK and second longitudinal have been positioned, and you are satisfied with their position, tie in any other remaining hull Longitudinals.

- At this junction, the framework is aligned and securely tack welded in position as previously described.

- If something does not look right, now is the time to correct any misalignments

Fitting the Keel

You will know when the Keel assembly, shown below, is proper positioned and aligned with the hull by the aligned relationship between the angle of the transom and the angle of the keel. The angle of the keel and the angle of the transom must be in plane from the bottom of the keel to the top of the transom in this design.

With the keel in place and welded to the structure the last two Longitudinals can be added to the framework and tack welded as previously described.

Notes

Plating
the
Hull

Plating the Hull:

Every Bezier True Rounds shell plate section, including 'Radius Chine' Single Constant Radius shell plate sections, are pre-engineered to fit an exact location on the hull.

For example, Shell plates in way of a Transverse frame include a curved line on the flat pattern. After the shell plate is formed, this line becomes straight, and is used as one of the references to align the shell plate to the hull.

Placing all the 'Primary' shell plate at their respective transverse frames takes out the cumulative error that could easily creep into the plating process, therefore these shell plates are designated primary.

To begin,

The photo shown on the next page represents a shell plate near the middle of the hull that is located on a Transverse frame. As previously described this shell plate has a line marked on the back that aligns with its transverse frame. Align these. Additionally align the notches on the long side of the shell plate with the Longitudinals frames.

Make a visual assessment by checking all the references discussed above. If the alignment is correct lightly tack weld the shell plating to the sheer longitudinal.

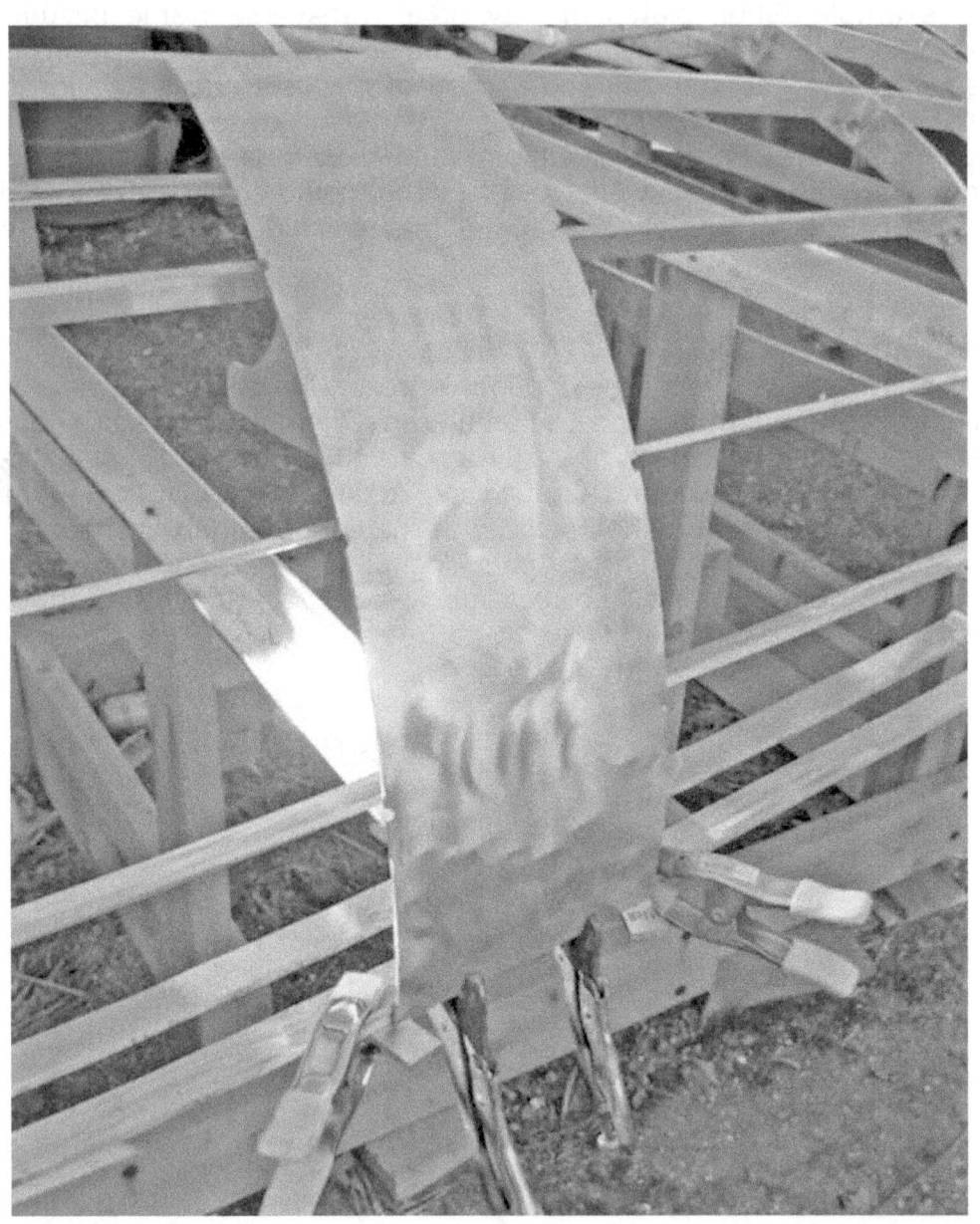

As seen, below, longitudinal notches in the shell plating are aligning nicely. Move the clamps to the next longitudinal. Check alignments before lightly tack welding the longitudinal at the half round notch to the shell plating.

In the picture, below, the clamps have been moved to the last mid longitudinal. It can be seen that the edge of the shell plating aligns perfectly with the 'Chine' longitudinal.

Fitting the Secondary Plating:

With the eight (8) primary sections in position the adjacent shell plates are ready to be fitted between them. These remaining sections will be worked randomly, alternating between sides of the hull.
The secondary shell plating is designed to fit against its adjoining segment. Any kerf required between segments for welding purposes is done after fitting the shell segment to the hull framework.

To begin,

The picture below shows the shell plate temporally clamped to the sheer longitudinal. Move and Rotate the shell plate about the clamp to obtain a fit along the girth of the plate.

Continue working the secondary shell plating to the framework as shown in the picture below, your main guide will be the edge of the adjacent shell segment and the longitudinal frame half-notches which align with the longitudinal frame.

Continue to work the shell plating toward the Chine Longitudinal as depicted, in the photo below, using the notches in the shell plating that align with the Longitudinal frames.

The accuracy of 'Bezier Chine Design and Construction' really shines in the below picture, where the shell plates lands precisely on the chine line longitudinal.

Below is the product of our work.

The same procedures is used to fit plating section between adjacent shell plates.
The shell plate below is shown loosely aligned by using two clamps. One clamps at the sheer line and the other at the chine line.

There, however, is a slight overlap that needs to be trimmed to allow the new shell plate to fit between the adjoining plating. Trim as required using a 'Cut-off' wheel.

The picture below shows the results of our work.

It is really that simple!

Notes

DEVELOPABLE
SHELL PLATING

Bottom Elementary Shell Plating:

- The bottom developable shell plating of the Prototype Bezier 12.5 was incorporated for ease of construction
- 1/8" thick hardboard strips four to six inches wide were used to template the Developable bottom surface of the hull.
- The bottom shell plating is divided into two pieces for ease of handling and material length considerations.
- For the stern template the hardboard strips overlap the transom & chine longitudinal by approximately one inch.
- Along the keel line the template extends under the keel shell plating.
- The seam line between plating segments will be four inches away from any transverse frame.
- With the hardboard template in place, on the backside, use a black marker to scribe the perimeter of the framework.
- On the front side, mark reference locations between the template and hull to realign the template back to the hull after it has been cut to shape.
- Place the template back on the hull, as shown in the photo below. Check for accuracy before committing the template to the actual shell material.

The plating for the aft section, shown below, is cut to fit under the keel side plating. Pressure between the two create a virtual clamp holding the shell plating to the curve of the keel along this edge. Use the reference marks previously described to align the shell plating back to the framework.

With more curvature in the forward shell plate, a little more coaxing is required to pull the plating to the framework. This section of shell plating is also set under the keel plating, giving a good foothold to draw the plating to the framework. Notice the three threaded rods, supported by transverse frames used to pull the plating to the framework.

In the picture below, the threaded rods are removed from the forward section of the shell plating.

Looking Inside the Hull:

All during the bottom shell plating process, the Builder was working entirely from the outside of the hull. This allowed the shell plating to find its own *'Natural Lay'*. There was no attempt to draw the shell plating to the transverse or longitudinal frames between the Cvk and Chine longitudinal. This developable surface plating procedure will always result in a *'Fair'* outside surface.

Looking under the hull, the 'Builder' is going to find that the shell plating is not in full contact at many locations with the transverse and interior Longitudinal frames, however it will be in full contact at the chine longitudinal, keel shell plating, and Cvk.

Floating the Longitudinals:

- Working under the hull the Builder is now going to break the tack welds loose between the transverse and longitudinal frames.

- The Builder will now move the Longitudinals to the shell plating. Hence the Term ***'Floating the Longitudinals'***.

- The Transverse frame in way of a developable surface are never welded to the shell plating. Only the longitudinal framing is welded to the shell plating.

- The diagram on the next page illustrates the adjustments to the framework that will be needed.

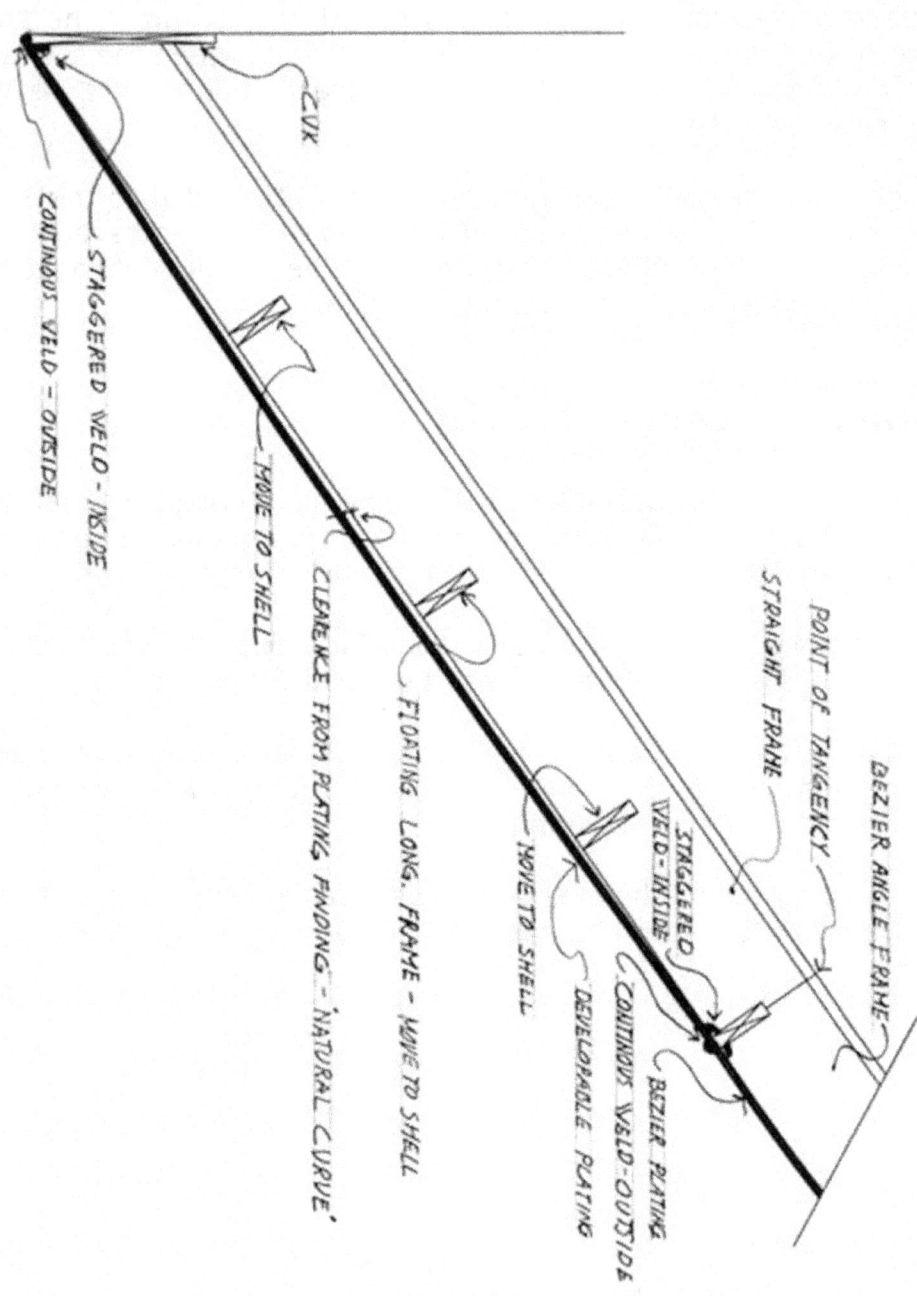

CVK

CONTINOUS WELD — OUTSIDE

STAGGERED WELD — INSIDE

MOVE TO SHELL

CLEARENCE FROM PLATING FINDING — 'NATURAL CURVE'

FLOATING LONG. FRAME — MOVE TO SHELL

MOVE TO SHELL

DEVELOPABLE PLATING

CONTINUOUS WELD — OUTSIDE

STAGGERED WELD — INSIDE

STRAIGHT FRAME

POINT OF TANGENCY

BEZIER ANGLE FRAME

BEZIER PLATING

Notes

Weld Lines

What is a Weld Lines:

Joining two (2) metal hull components together by welding, is referred to as a **'Weld Lines'**. A Weld Line never joins more than two hull components at a time

When a Weld Line joins more than two components a Tri-Axial stressed weld seam is created causing undo internal stress in the weld line. Tri-Axial stress joints are to be avoided.

A Tri-Axial weld joint is illustrated below, and is recognize by the convergence of three hull components joined by a single weld line. In the illustration below, two (2) shell plates and a Transverse frame make up a Weld line.

This tri-axial weld line is easily avoided by running the shell plating a calculated distance past any transverse frame, as will be shown later.

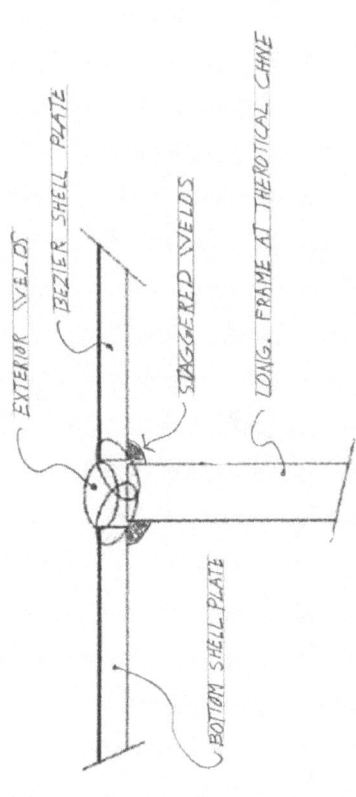

A Tri-axial weld joint is also formed when Weld line intersect or cross each other. The below sketch illustrates a Tri-axial created by the weld lines of the CVK, Transverse Frame, and Shell Plating.

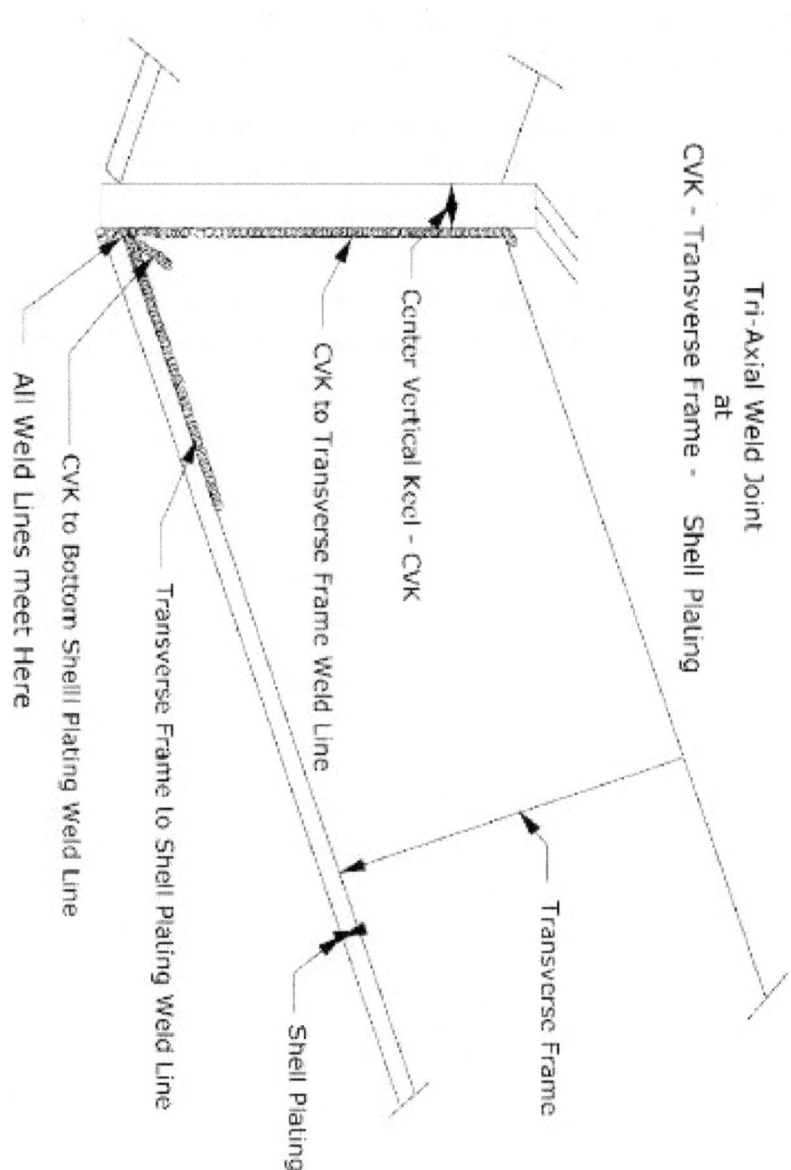

Tri-Axial Weld Joint
at
CVK - Transverse Frame - Shell Plating

Center Vertical Keel - CVK

CVK to Transverse Frame Weld Line

Transverse Frame to Shell Plating Weld Line

CVK to Bottom Shell Plating Weld Line

All Weld Lines meet Here

Transverse Frame

Shell Plating

In the below illustration you can see that a simple **Snipe** in the Transverse frames separates all the Weld Lines, thereby eliminating the tri-axial stress weld line shown in the previous illustration.
Note: That the **Snipe** also acts as a **Limber Hole** allowing water to pass between the Transverse frames to the lowest point of the bilge.

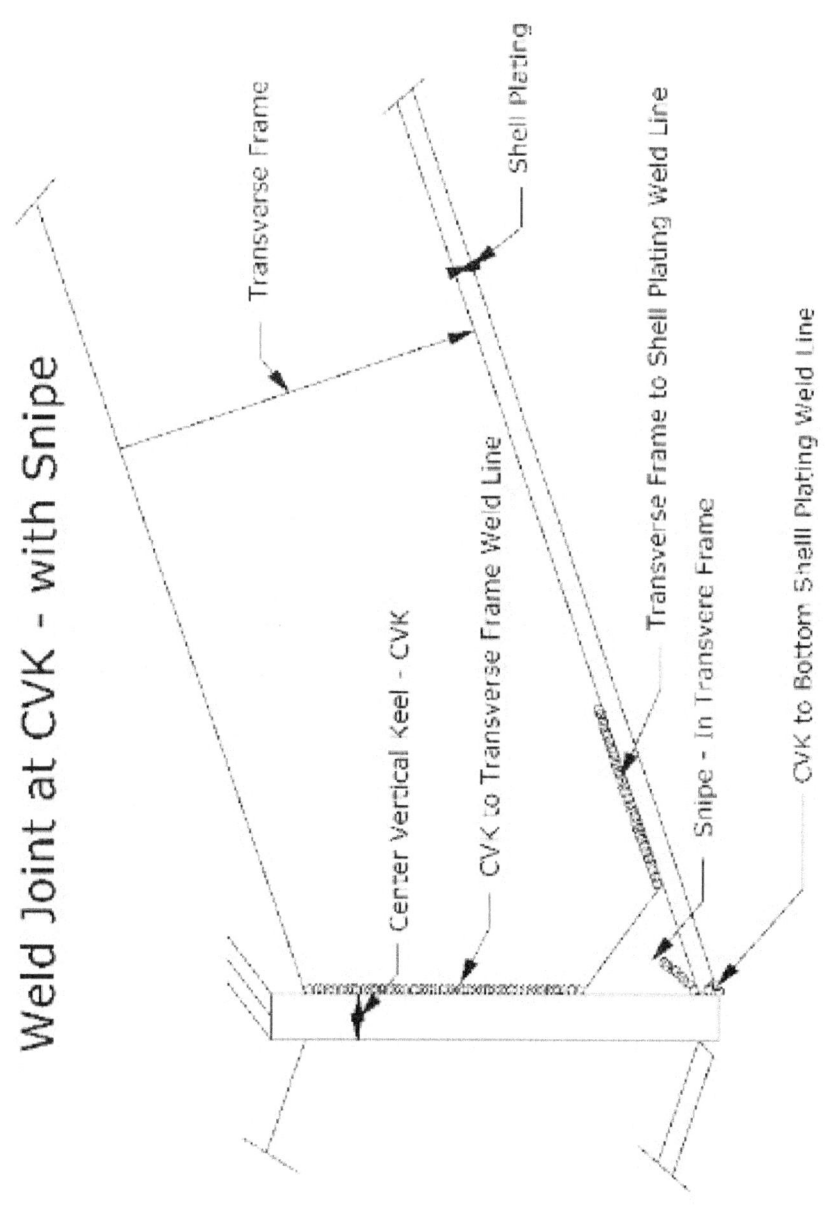

Weld Lines
Between Shell Plating and longitudinal frames:

Place a **Rat Hole** in the longintudinal framework where every shell plating crosses the longitudinal frames as shown in the below sketch.

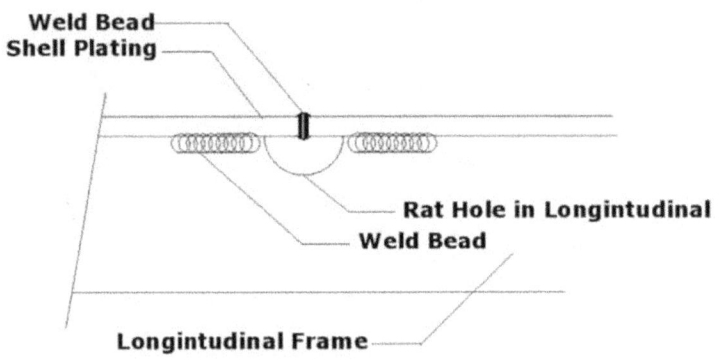

Weld Lines
Between Transverse Frames and longitudinal frames:

Here the Transverse is notched to accept the longitudinal frame. Weld around the entire notch in the transverse frame. Remember that the hulls framework is welded before the shell plating is applied.

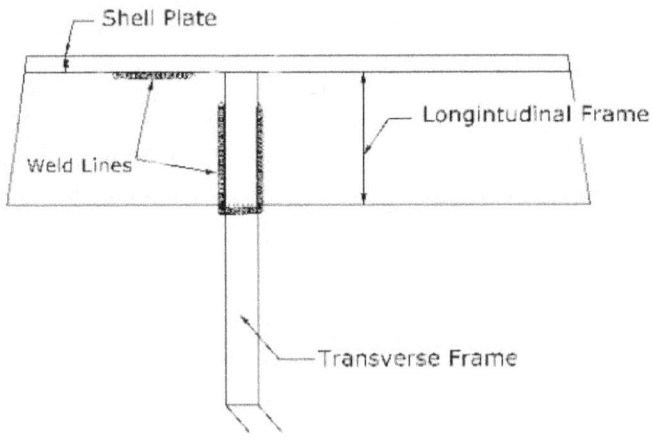

Weld Lines
Between Shell Plates:

After the shell plating is fitted to the hull framework, it is customary to weld the inside of aluminum shell plating before the outside. This allows the outside of the shell plating to be easily back-chipped to remove impurities from the weld by grinding, chipping, or wire brushing before welding the outside of the hull.   I suggest Appling 'One-Sided' welding tape to the outside of the hull to prevent the atmosphere from contaminating the open root of the Weld line on the outside of the hull shell plating thus reducing the amount of grinding chipping, and wire brushing in preparation for welding the outside of the shell plating.

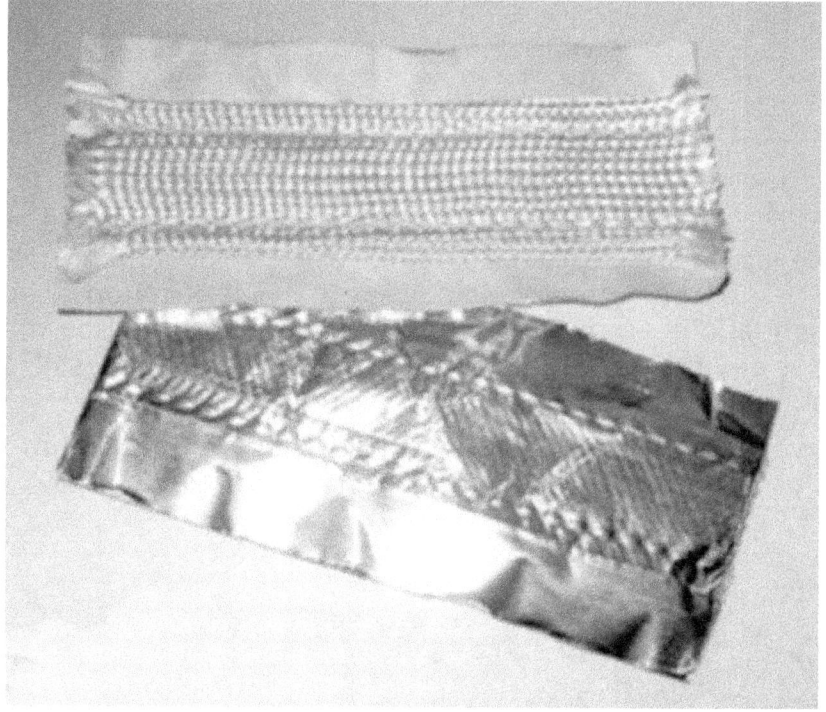

Refer back to illustration where the shell plating crosses the Transverse and longintudinal frames.

Weld Lines
Shell Plate to Watertight Bulkheads:

A small open aluminum keel-boat such as the Bezier 12.5 requires flotation in the event of swamping. To this end the boats that I design with watertight tight bulkheads and internal tanks will be fabricated using the techniques described below. Used to illustrate will be the 'Watertight Bulkheads used to create flotation void spaces in the Bezier 12.5. Keeping in mind that Classification society's do not consider voids in themselves to be floatation unless the void contains flotation material.

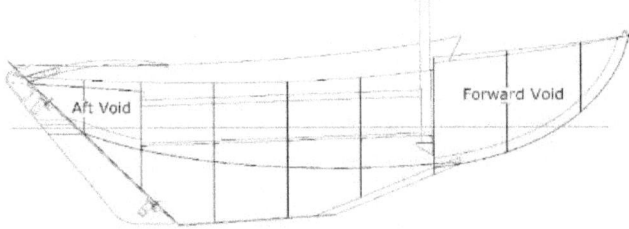

It will be up to the Builder to make the call on adding flotation material to be in compliance with Classification Society's recommendations. Knowing that voids filled with flotation material will hinder any future work in these areas and will be messy, to say the least.

The Forward Void is enclosed by a solid sheet of aluminum which doubles as Transverse Frame #3. In the half drawing below:

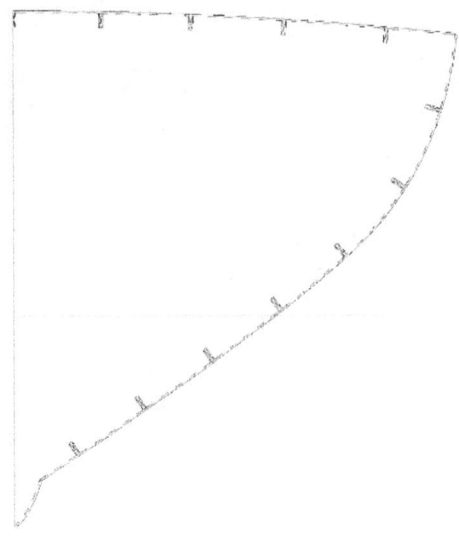

- This Watertight bulkhead will act as a Transverse frame.
- The dash line indicates the inside of the shell Plating.
- The solid line indicates the outside surface of the Shell Plating and the outside edge of the watertight bulkhead. In other words, the Watertight bulkheads penetrates the shell and deck plating and are flush with the outside surface of each. plating.
- Where the watertight bulkhead crosses the hull and deck longitudinal frames are welded as any other Transverse frame.
- The Watertight bulkheads itself will be fitted with a watertight access cover plate, (not Shown).

The same 'Weld Line' configuration as previously describe for the joining of the 'CVK' to the Shell Plating at the bottom centerline of the hull.   The difference is that the 'Weld Line' run in Girth.

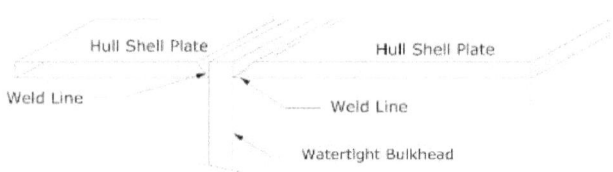

The Void bulkhead 'Weld Line' at the deck is similar to the 'Weld Line' joining the Shell Plating.

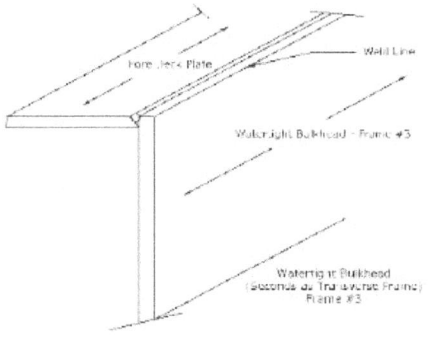

The proceeding situations are another place where 'One-Sided' welding tape makes the construction process manageable.

Where the Watertight Bulkhead penetrates to the outside of the shell plating place the 'One-Sided' weld tape one the inside corner between the shell Plating and Watertight Bulkhead. In this case we will be doing a 'One-Sided' weld, since it would be virtually impossible to weld inside the void.

Where the Watertight Bulkheads meet the Fore Deck plating again place the 'One-Sided' weld tape on the inside corner between the Fore Deck and the Watertight Bulkhead.

Weld Lines
Keel Structure:

Note the **'Rat Hole'** in the Transverse frames, in way of the keel structure which separates the three (3) 'Weld Lines' that attach the keel to the hull.

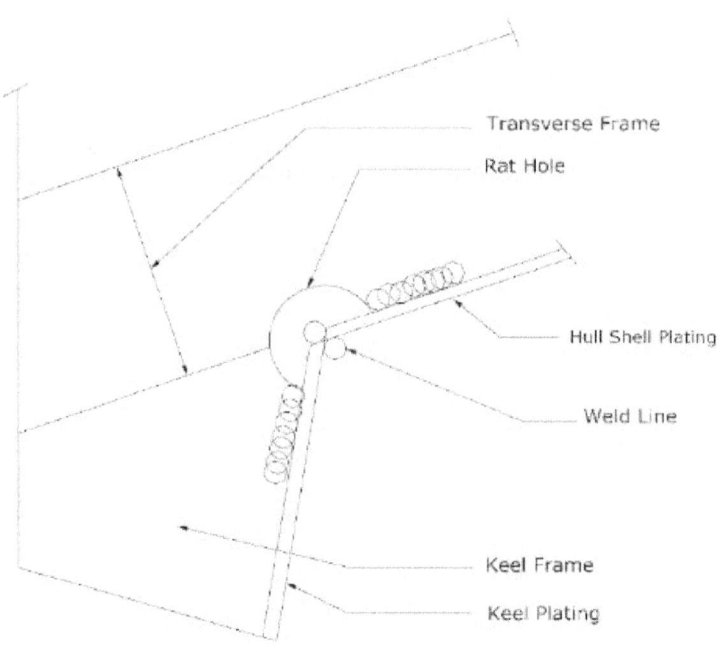

The following illustration shows how the nose edge of a typical Finn keel is formed by a specified diameter pipe. There are two Weld line, one outside on the keel surface – the other inside the hollow keel structure. Note how the keel shell plating flows tangent and smoothly into the leading edge.

Note, the 'Snipe' at the bottom edge of the keels Transverse frame where the Keel Shoe meet the Keels side shell Plating. This is another place that the Builder would use 'One-Sided Weld Tape.'

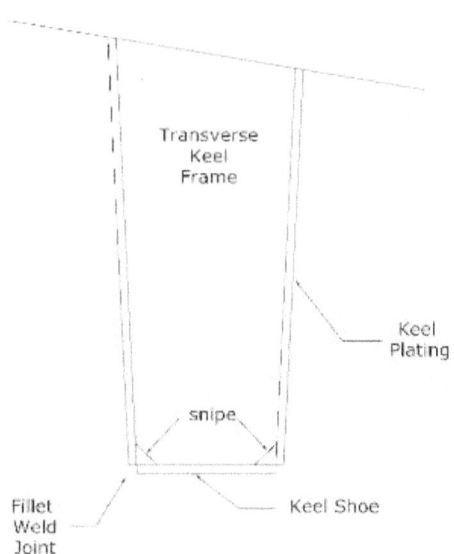

The trailing edge of the keel has the same 'Fillet Weld Joint' as the above Bottom shoe. Again, another place where 'One-Sided' welding tape would come in handy

Plug Welding:

It is often impossible or impracticable to weld inside a narrow structure such as a sailboat keel, especially when fabricated from aluminum do to the size of aluminum welding torches.

The following sketch shows a section view of a keels transverse frame where the frame will be Plugged welded to the shell plating of the keel.

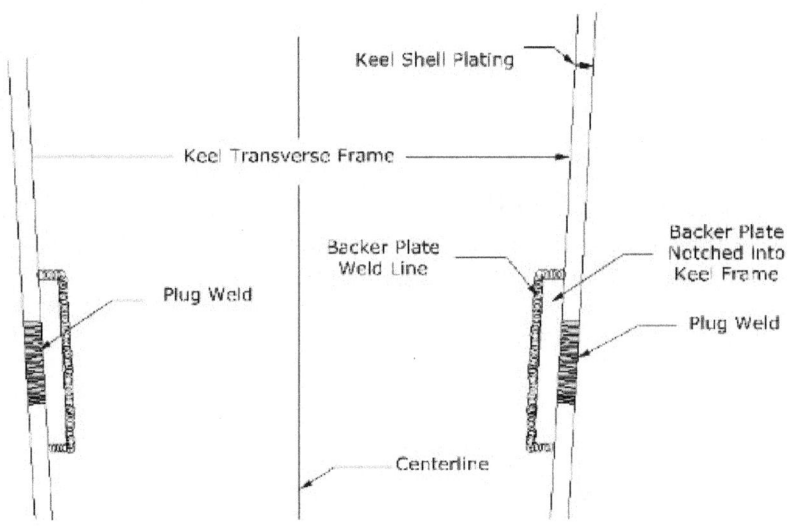

Note the 'Backer Plate' notched and welded to the keels transverse frame on both side of the keels transverse frame. It is at this location where an obround slot will cut into the keels shell plating.

The following illustrations shows the location of the keels transverse frames and obround slots.

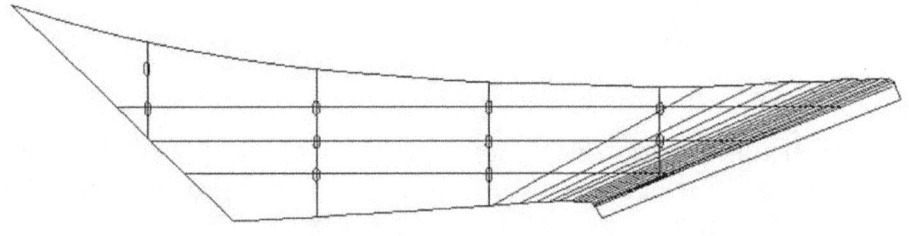

Continuous Welding Sequence:

- Before any Continuous Welding commences, the whole of the hull structure must be tack welded inside and out.
- The general rule for aluminum construction is to weld the inside of the shell plating, with 'One-sided' welding tape applied to the outside surface. This allows the outside of the shell plating to be easily back-chipped to remove impurities from welds by grinding, chipping, or wire brushing.
- For steel construction I always welded the outside of the hull first. I really believe that it all boils down to 'Which is easier'.
- Whatever two component of the hull is being fully welded the key is to skip around or back-step, using short welds.

For example:

- When welding the outside seams of the true round hull plating, start near the middle of the hull, and run a two-inch weld. Then, move to the same location on the opposite side of the hull and run a two-inch weld there.
- Next, change locations to another plating seam, working forward or aft from the first weld.
- Repeat this procedure over and over until all the true round plating seams have been welded.
- Use this method for welding shell to shell plating, longitudinal framing to shell plating, keel platting to shell plating, etc.

General Thoughts:

- An 'Aluminum' hull is a totally separate animal from a 'Steel' hull. Aluminum requires particular attention to how its components are deburred, meet, and welded. Remove all burrs and sharp edges.
- Different grades of aluminum require certain bending radius when forming in a press-brake. Certain aluminum grades such as 6061 can only be bent across the grain of the material, never with the grain.
- Do not Grind the welds on the outside of the shell plating until the inside of the has be completely welded.
- Never weld any section of the True Round or Developable shell plating to any transverse frame.

Notes

Surface
Finishing

Surface Finishing

No matter what type of metal hull you build, grinding and sanding weld seams on the outside surfaces are always a given.

- The hull type with the least footage of weld is a Single Hard Chine design.
- I think a Double Hard Chine would be next.
- Multi-Chine would follow that.
- Bezier' True Round designs and 'Radius Chine' designs would have the most weld footage.

So, what are you going to build, 'The quick and easy' Single Hard Chine design or 'The not so quick but easy' Bezier' True Round or 'Radius Chine Design.

Before you decide, think about how you are going to look at your Single Hard chine boat that was so 'Quick and Easy' to weld over the True Round design that you really wanted, years later. ***What will the time that you initially save mean now!***

Flushing the Welds to the Hull Surface

The weld seams, shown below, between the True Round sections of shell plating need to be sanded flush.

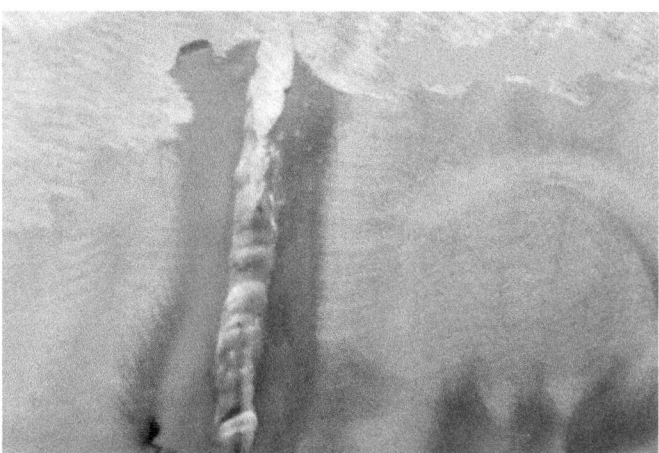

When sanding any weld seam flush, we can expect two different results. The first result shows the weld material evenly removed. Care was taken to ensure that there was no reduction of shell thickness in this process.

In the photo below, hollows can be seen after the weld was sanded flush between adjacent surfaces. Further grinding to remove these hollows would sacrifice hull thickness. It is not advisable to continue sanding any further. Superficial hollows such as these can be clad welded and sanded again or filled with 'Fairing' compound.

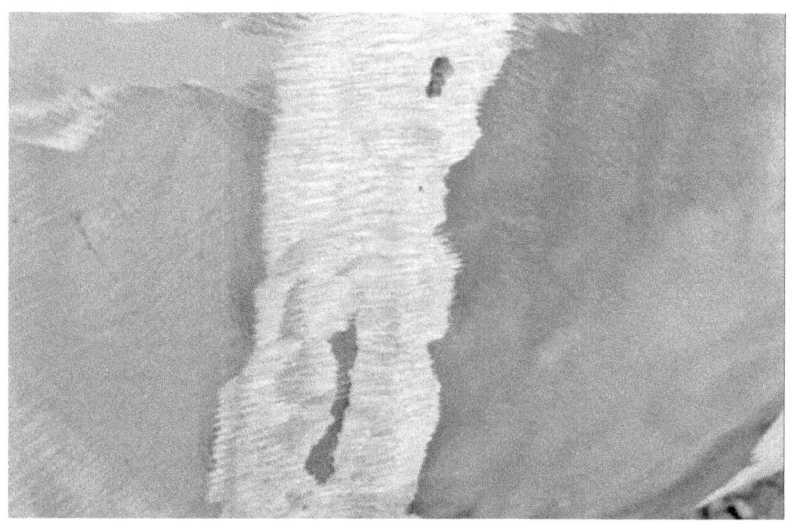

The process of initially leveling the weld seams leaves visible sanding marks. The five seams on the left side needs to be refined further. The surface on the right side has been refined by an additional sanded with a fine grit paper, then the surface was than brushed with a light grade rotary power stainless steel wire brush.

Any further sanding would only compromise the thickness of the shell plating. The surface of the hull felt and looked fair to my hand and eye. I did consider stopping here and painting the hull.

However, Common sense intervened. This was the Prototype and 'Proof of Concept' for 'Bezier Chine Design and Construction'. I would have the hull professional Faired.

Fairing the Hull Surface

- From my observations, of the Fairing process:
- The surface of the hull was washed with an aluminum surface solvent.
- The hull was than lightly sprayed with a grey paint.
- After the paint dried the hull was 'Longboarded'.
- In the following pic, the section of the hull that are darker in color are the low spots.
- A thin layer of Fairing compound was applied, I would assume in some measured way, and longboarded again.
- The hull was lightly sprayed painted again. Then longboarded again.

Several iterations of the above process resulted in the photo below. The hull was Spray painted "Cloud White". A two-part epoxy paint formulated for the fairing system compound that was used.

The **'Bezier 12.5'** is as fair as any production fiberglass hull.

Notes

Deck
And
Cockpit

Deck Framing

The photos below show the bow and aft deck Longitudinals and the Cockpit Coaming which doubles as a structural member that ties all the Transvers frames at the sheer-line together. Additionally, the Cockpit Coaming serves as the seat back.

Decking

Aluminum decking is used for the breast plate between the fore point and frame #1, the fore deck section inside the coaming, and the narrow side decks aft of frame #3 to the transom.
Marine plywood decking is used for the fore and aft decks. The fore deck, shown below, is shown being templated.

Plywood decking, shown below, is secured to the framework around its perimeter by flat head stainless steel machine screws approximately 4" apart. The screws are hidden by the deck trim.

Bulkheads

Void spaces are located under the fore deck forward of Frame #3 and the aft deck aft of Frame #7 and are used as flotation compartments. Bulkheads of 3/8" plywood enclose these areas. The aft void is completely filled with 9 square feet of spray in close cell flotation. The forward void was semi-filled with 18 square feet of flotation, a total of 27 square feet of flotation. Minimum flotation is 18 square feet.

The aft bulkhead shown in the below pic is secured to the transverse frame by #10 stainless steel machine screws.

Cockpit Ceiling

Perforated 0.040" aluminum, shown in the below pic, is used for the 'Ceiling' material that conceals the cockpits structural members. It is purely cosmetic. Wood cover strips 5/16" by 2" wide conceal the seams between the ceiling sections at each transverse frame. The wood strips are secured to the transverse frames by stainless steel machine screws drilled and tapped into the angle transverse frames.

Added support, shown below, for the ceiling between the transverse frames is provided by clips attached to the longitudinal framework.

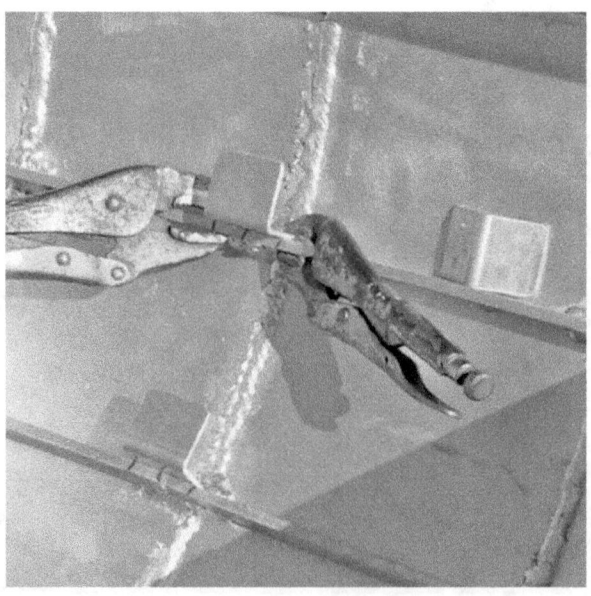

Cockpit Seats

A template is resting on the flat bar seating frames four, five, and six and is fashioned to the desired shape of the seats.

Below is the seating cut to shape using the hardboard template. A 3/16" space was left between the boards. The seating boards are tied together by cross members secured by screws from the bottom.

The cockpits seats, shown below, are supported by the hulls Transverse framing system.

Cockpit Sole

The cockpit sole was templated and constructed similar to the seating. The sole consists of seven (7) boards, joined to create three (3) sections that consist of a center board and left/right end sections.

The cockpit of the Bezier 12.5.

Rudder

The rudder was shaped from a single piece of mahogany 12" wide and 1 1/2" thick. It is profiled from the supplied full-size pattern. The two upper lines mark the boot stripe, the third line indicates the designed waterline and the transition lines represent where the foil shape below the designed waterline returns to the upper stock thickness of 1 1/2". The picture below, show the rudder hung off the transom.

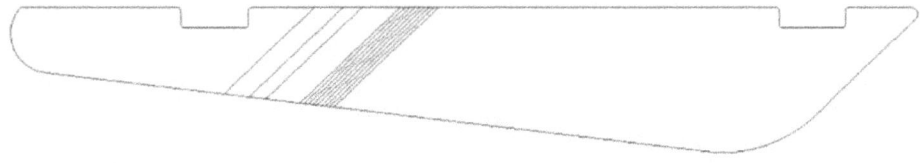

Notes

Theory
of
Approximate
Development

In sheet-metal work, surfaces are divided into two general classes: 'Elementary' and 'Warped'.

Elementary Surfaces:

Elementary surfaces can be developed accurately onto a single plane and formed back to their original three-dimensional form by simple folding and rolling techniques without stretching or compressing the material.

Elementary Surfaces are surfaces that:

- Lie on a Single Plane

- Curved Surfaces or sections of a curved surfaces.

- Radial Surfaces or sections of a radial surfaces.

An example of a Plane Surface is a rectangular box. Here random lines can be drawn anywhere on any side of the box. Lines such as these are designated as 'Elements' of the surface, because they lie in full contact with the surface.

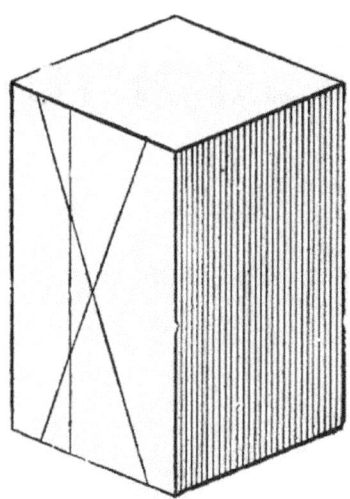

A Curved surface is a surface where no three consecutive parallel 'Elements' are on the same plane. To be an 'Elements' of a curved surface it is required to be in full contact with the surface.

A Radial surface is again a surface where no three consecutive radial 'Elements' lie in the same plane. To be an 'Elements' of a surface radiating from an apex it is required to be in full contact with the surface.

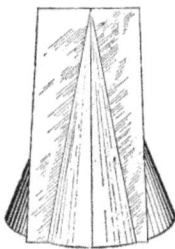

Any Sheetmetal or Plate fabrication composed of any combinations of these surfaces can be developed accurately onto a single plane. This includes such 'Fabrications' as Single, Double, and Multi-Chined metal and plywood boat hulls.

For sheet-metal and plate objects consisting of all Plane surfaces mathematical formula's such as **'Bend Allowance'** *and* **'Bend Deduction'** are used to 'Unfold' the fabrication onto a single Plane. Full details can be found in my book, ***'Applied Metal Boatbuilding Methods - Sheetmetal Pattern Development'.***
For sheet-metal and plate objects that are a combination of Elementary Planes, Curves, and Radial 'Elements' layout methods know as Triangulation, Parallel Line development, and Radial Line development are used to 'Unfold' the fabrication onto a single Plane.

Warped Surfaces

A warped surface has no 'Elements'. It is a surface where a straight edge makes contact at a single point only. A Sphere is a good example of a surface that cannot be 'Unfolded' 100% accurately onto a single plane. However, such a surface can be developed 'Approximately' to lie on a single plane.

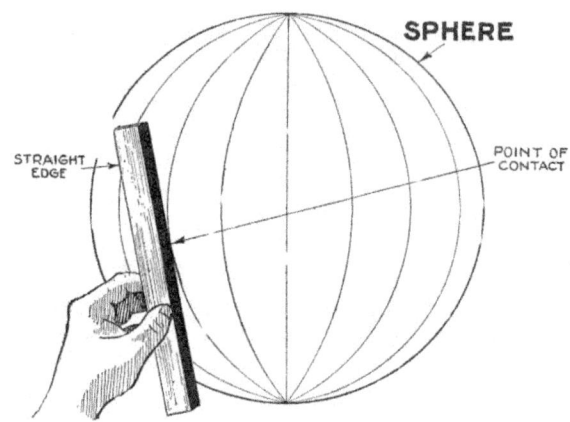

The Sphere is a good place to begin to understand how 'Approximate' development works, since it is a symmetrical object where only one pattern is required as compared to the many patterns required to 'Unfold' the warped surface of a true round metal boat hull.

Since 'Approximate' can be a relative term, a plan view of two (2) twelve-inch (12") spheres are shown cut at their diameters. One Sphere will be divided into eight (8) chords-segments and the other into sixteen (16) chords-segments. The chord line represent the segment of the Sheetmetal sphere developed 'Approximately' onto a single plane.

We could have divided the sphere into even more chords-segments, say thirty-two (32) or even sixty-four (64) or any amount we desire. Never the less the eight (8) and sixteen (16) chord arrangements will adequately describe the theory of 'Approximate' development for Warped or Compound surfaces.

The below illustrations shows the Eight (8) 'Approximately' developed chords-segment layout 'Unfolded' onto a single plane using 'Parallel Line Development'.

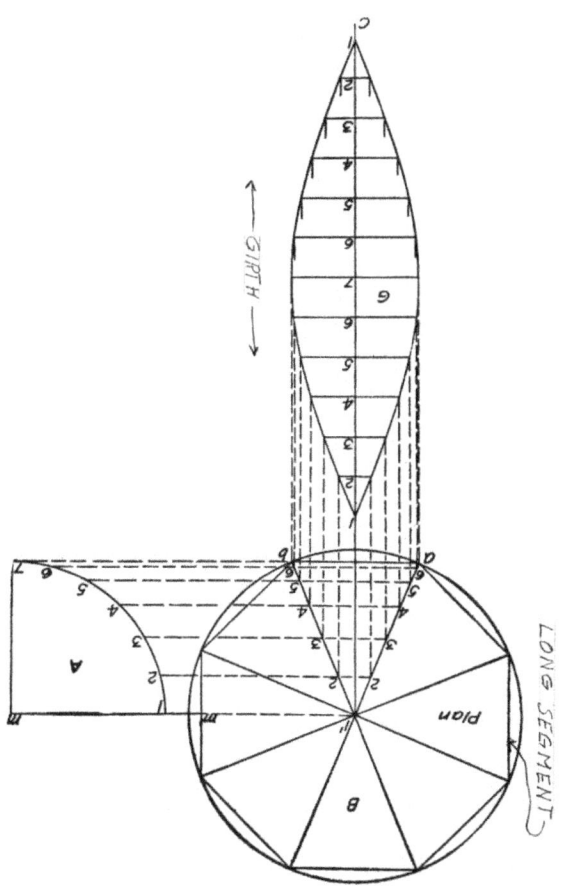

In the eight (8) segment layout, the chord length is 4.592", the distance between the chord and the finished surface arc of the sphere is 0.456", and the arc length of the finished surface of the sphere is 4.711". The difference in length between the chord length and the arc length is 0.199". The sphere therefore has been 'Approximately' developed. When assembled the eight (8) segment sphere would be very prismatic in form.

The sixteen (16) segment sphere, however, would be less prismatic in form. In the sixteen (16) segment layout the chord length is 2.341", the distance between the chord and the finished surface arc of the sphere is 0.115", while the arc length at the finish surface of the sphere is 2.356". The difference in length between the chord length and the arc length is now 0.015".

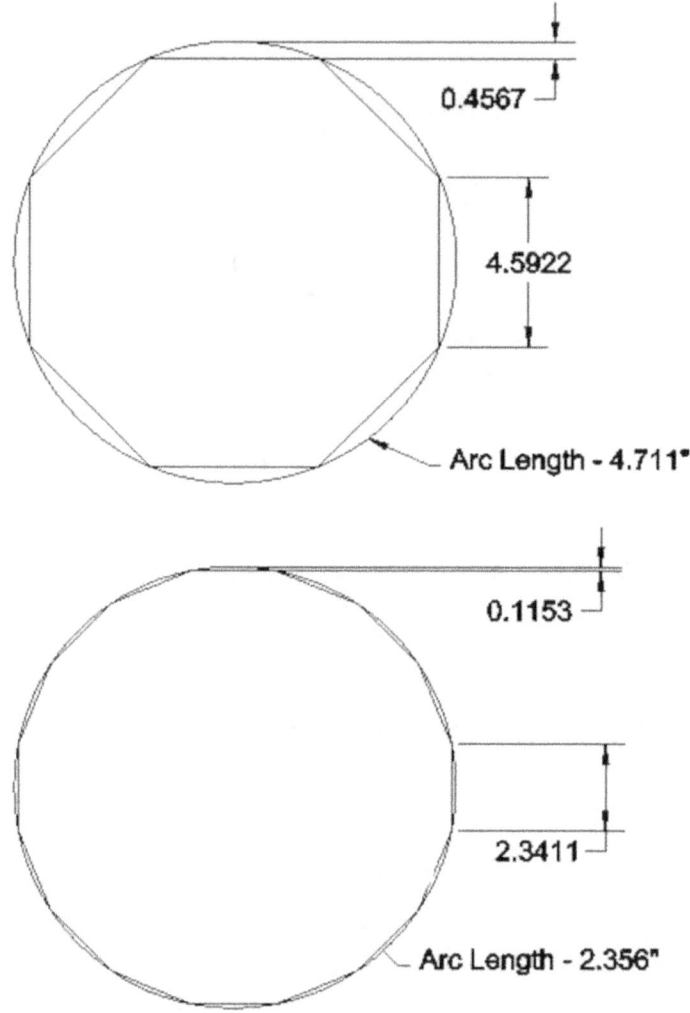

0.4567

4.5922

Arc Length - 4.711"

0.1153

2.3411

Arc Length - 2.356"

The 'End User' could accept the eight-sided sphere for the top of their roof final or they could opt for the sixteen-side version which brings the fabrication closer to the true surface of the sphere.

If neither of the above were acceptable a peening or hammering process would be required to stretch the material to its true shape as seen below. It should be apparent that the eight (8) segment version would require more peening, but have less seams, while the (16) segment version would require less peening, than again there are more seams.

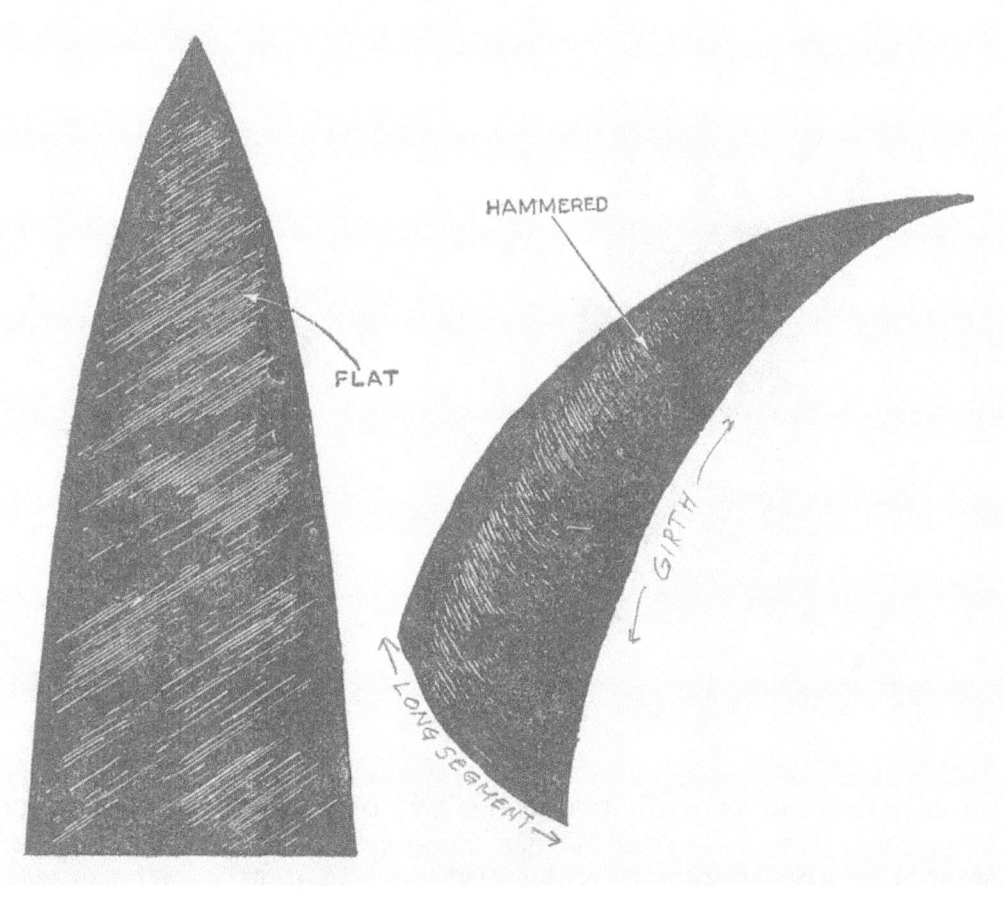

FLAT

HAMMERED

GIRTH

LONG SEGMENT →

Approximate Development of Shell Plating:

Before proceeding to 'Unfolding' the Warped surfaces of a True Round Metal boat hull, we need to relate the terms 'Girth' and 'Length' used in hull design to the Sphere below to relate the two vastly different 'Fabrications' to each other

.

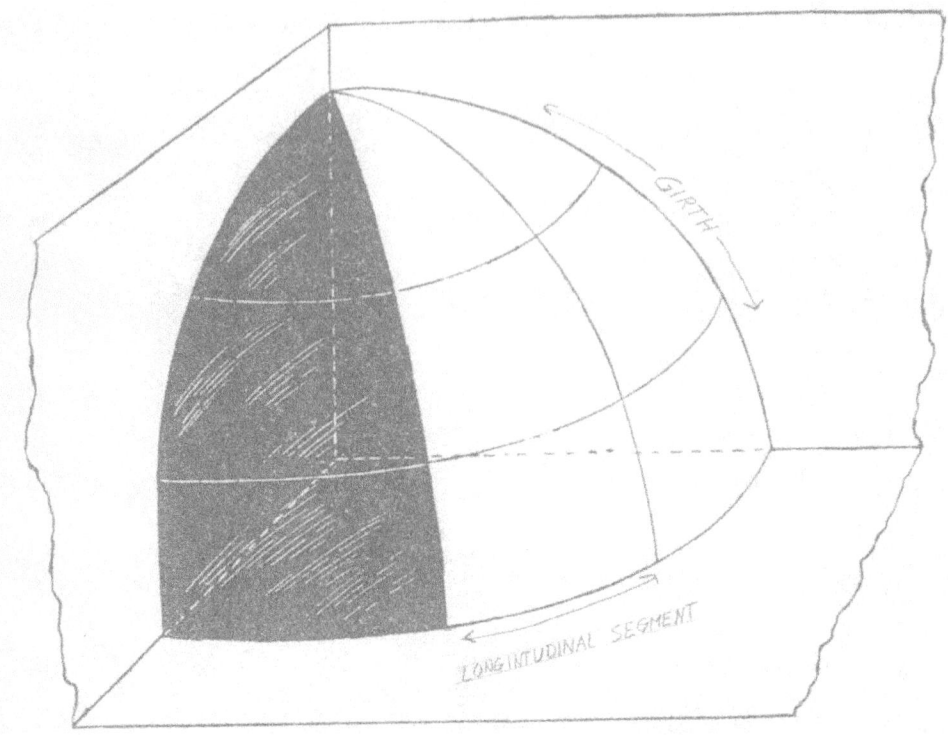

The only real difference between the two objects is that the Sphere needs only a single 'Approximately' developed pattern to define every segment of the sphere in both 'Girth' and 'Length'. Whereas the numerous segments that makeup the surface of a true round hull need to be developed individually since the 'Girth' and 'Length' vary at every longitudinal segment of the hulls shell plating.

Below the surface of the Bezier 12.5 has been divided into twenty-three (23) sub-surface segments along the length of the hull, which corresponds to the segments use to divide the Sphere.
For clarity, only the sub-surface segments that coincide with a transverse frame are shown. There are however two sub-surface plating segments between the one's shown in the below illustration.

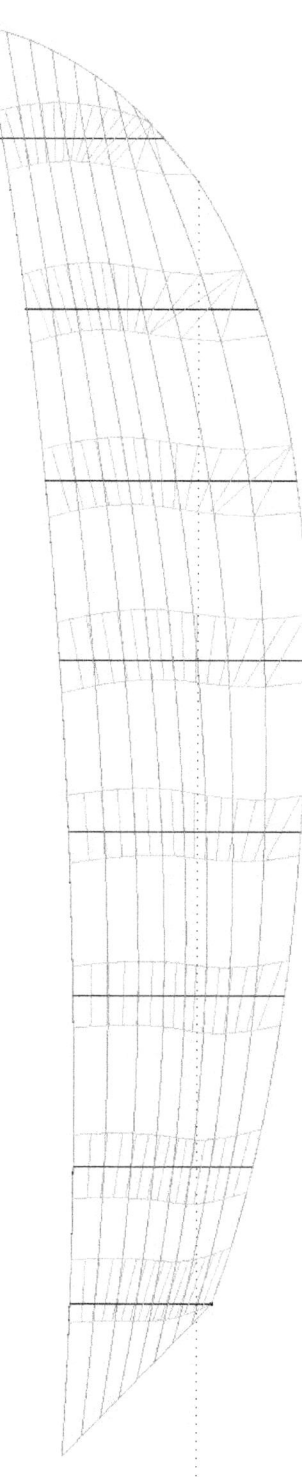

Longitudinal Direction of Roll:

Just as the Sphere was divided into chords at its diameter the free-formed curve of the Sheer-Line, shown in 'Plan View', is used to illustrate the chords along the length of the hull. The freeform curve represents the true form of the 'sheer-line', while the chord segments represent the 'Approximately' developed prismatic 'Sheer-line' intrinsic to 'Approximate Metal Fabrication'. The nodes define each segmented surface.

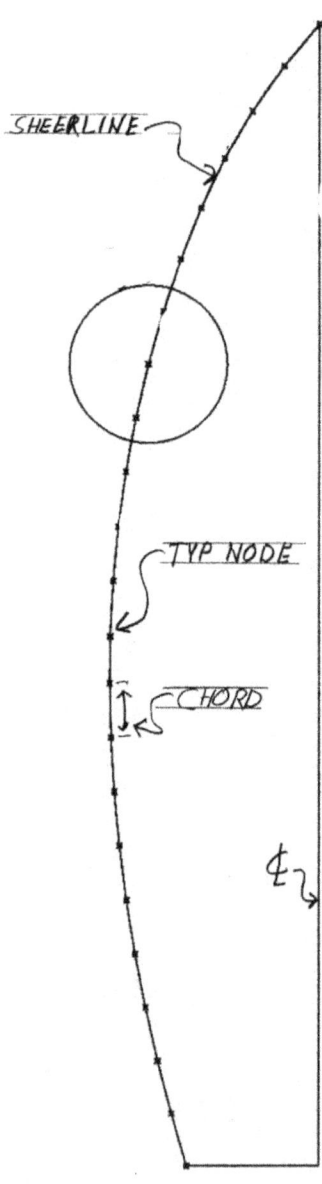

The below drawing illustrates a single line view between the free-formed sheerline and the chorded sheerline. Here the length of the chord is 8.681",
while the length of the free-formed sheerline between the chords endpoints is 8.687". The difference being 0.006". The distance at the centerline of the chord to the free-formed 'Sheer-line' is 0.047".

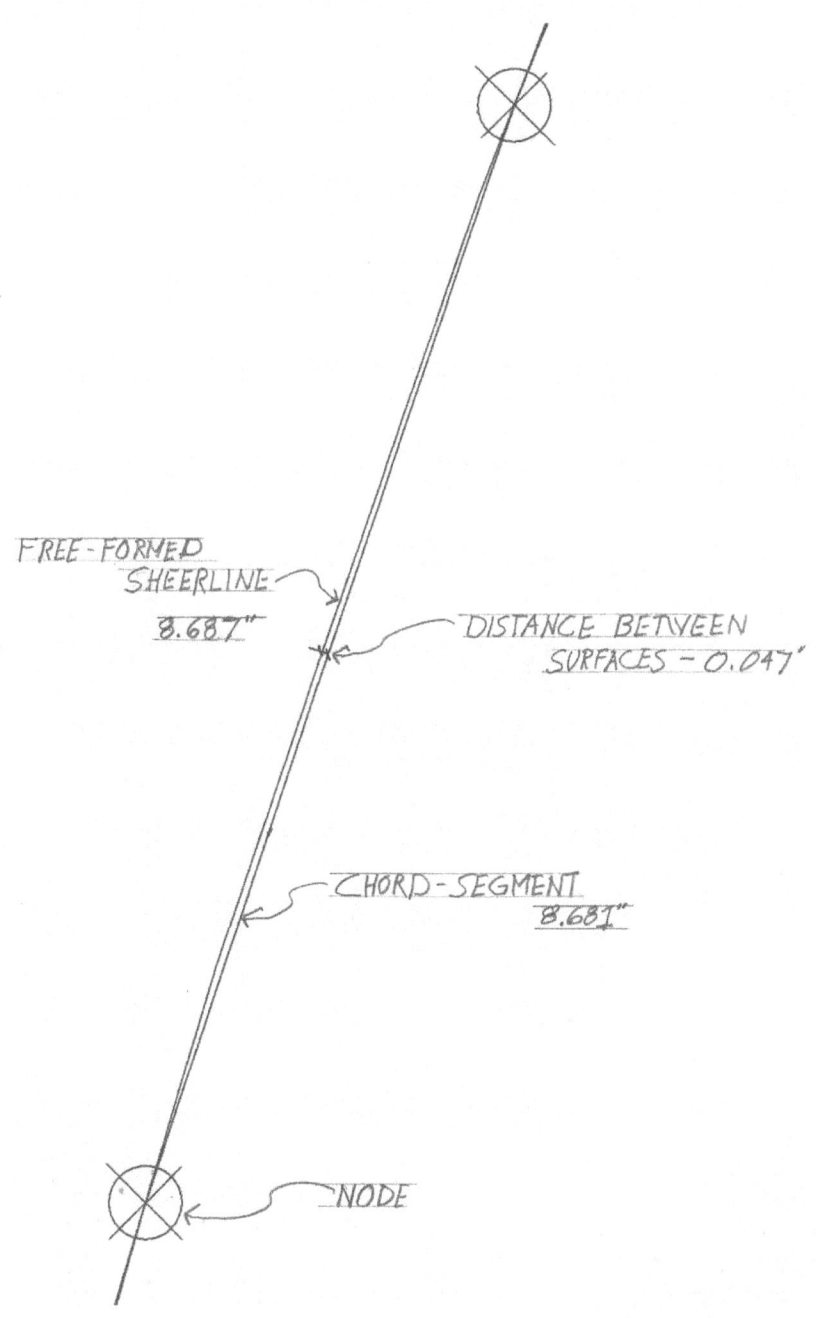

FREE-FORMED
SHEERLINE
8.687"

DISTANCE BETWEEN
SURFACES - 0.047"

CHORD-SEGMENT
8.681"

NODE

The following overall illustration, at the sheer-line, shows the relationship between the Chorded shell plates which were calculate by the process of 'Approximate Development' to the Compound curved surface from which the chorded shell plates were derived.

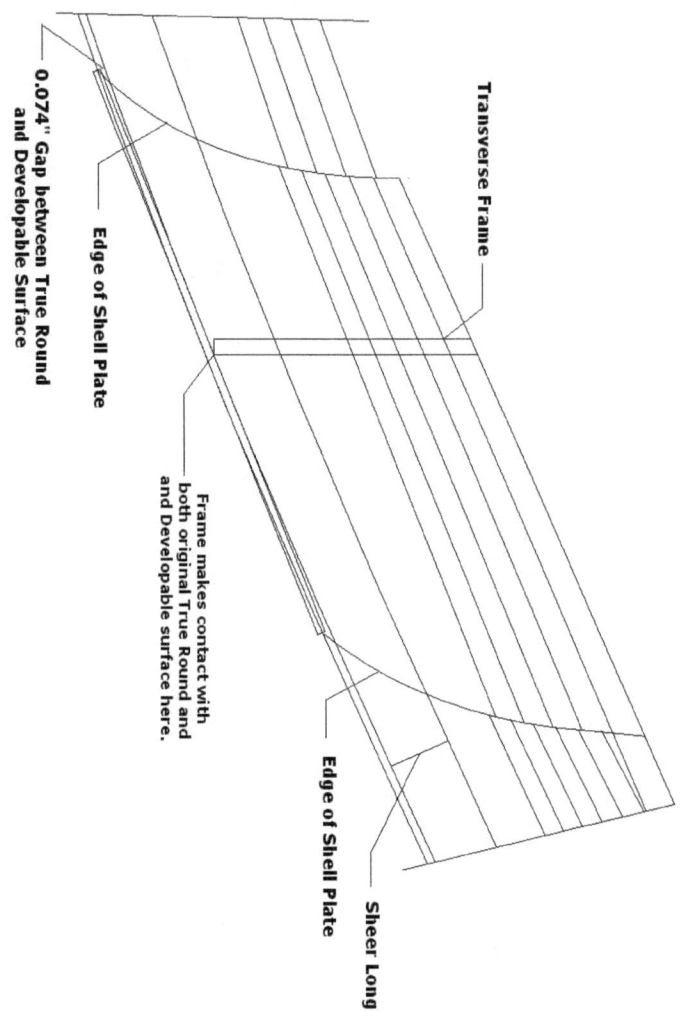

The following detailed illustration shows a closeup view of one-half of the above Chord segment. Notice that the segmented surface shell plate, unsegmented hull surface, transverse frame, and longitudinal all fall at a single location.

Also notice, while the designed True Round hull surface and the longitudinal frame both curve inward, while the segmented shell plate surface run straight out.

Here is where the 'Approximate' in 'Approximate Development' revels itself at the ends of the Chorded segment. In this case there will be a 0.074' standoff between the longitudinal frame and the chord segment shell plate that needs to be drawn together.

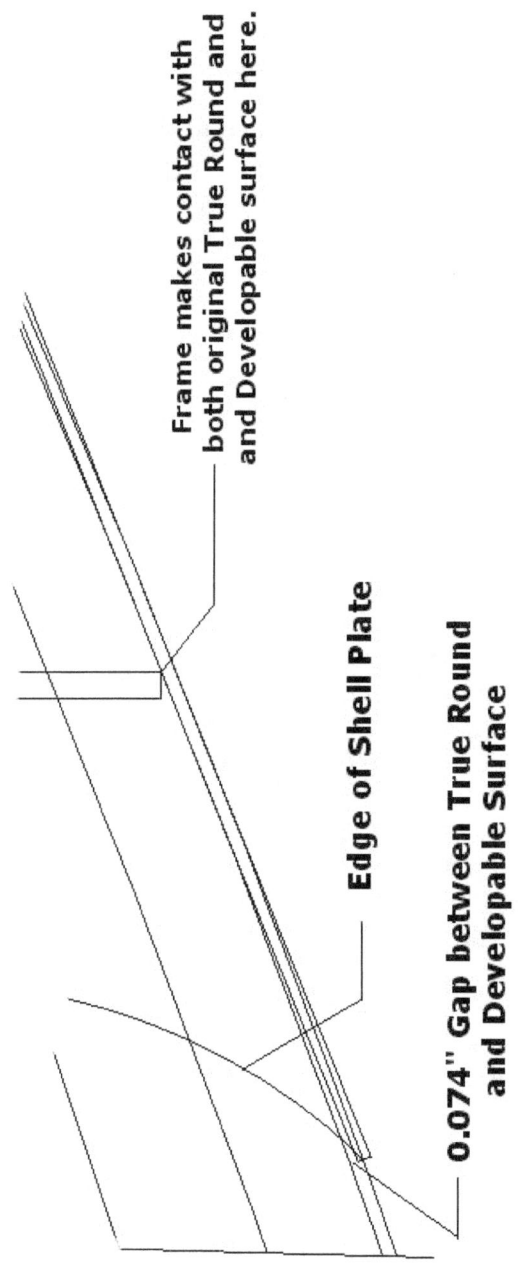

Frame makes contact with both original True Round and and Developable surface here.

Edge of Shell Plate

0.074" Gap between True Round and Developable Surface

The Roll in Girth:

Dividing the length of the hull into segmented surface Chords is one direction of roll. Dividing those segmented surfaces into chords become the other direction of roll.

A 3D view of the segmented surface at Transverse Frame Three is illustrated below. The horizonal or nearly horizonal line represent three-degree (3) Bend Lines. They are the 'Elements' of the surface making the space between these lines, Chords.

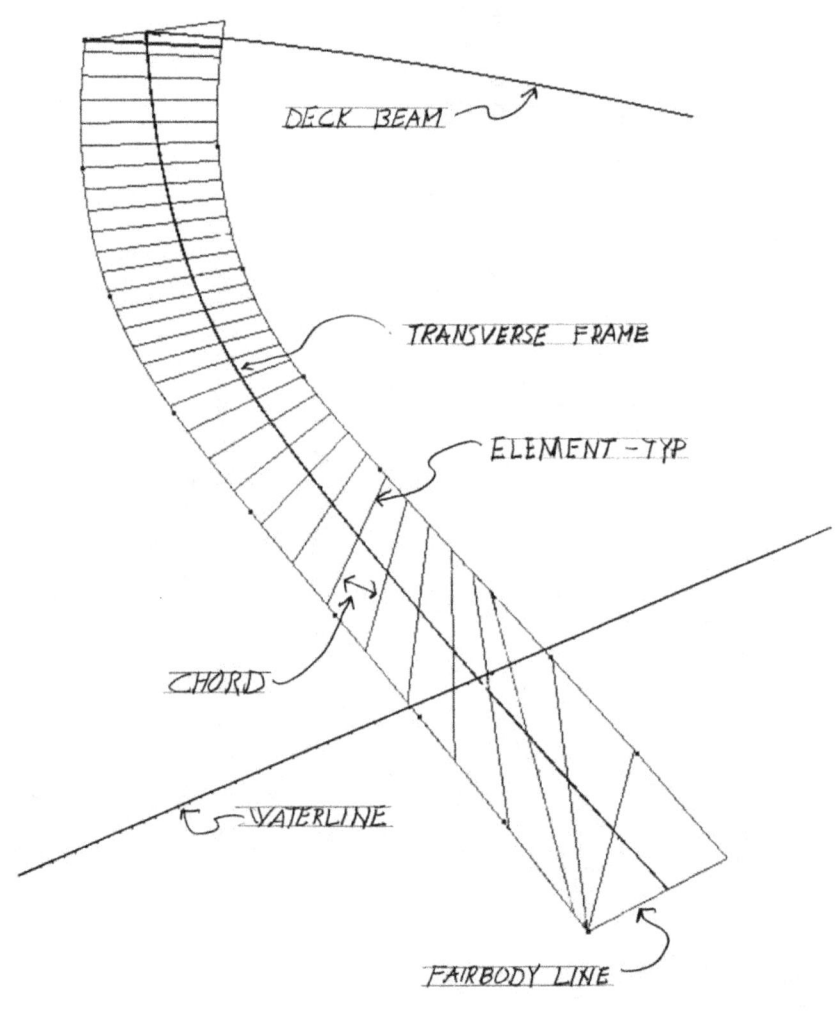

Notes

Other
Hull Configurations

Single Surface Bezier Design:

Bezier single surface metal hull designs are designed are true round free-formed designs just like and production fiberglass design where the surface runs from the Sheer Line around the turn of the bilge to the Fairbody Line at the bottom centerline of the hull. Shown in the below drawing are the Transverse Frames and the Design Waterline.

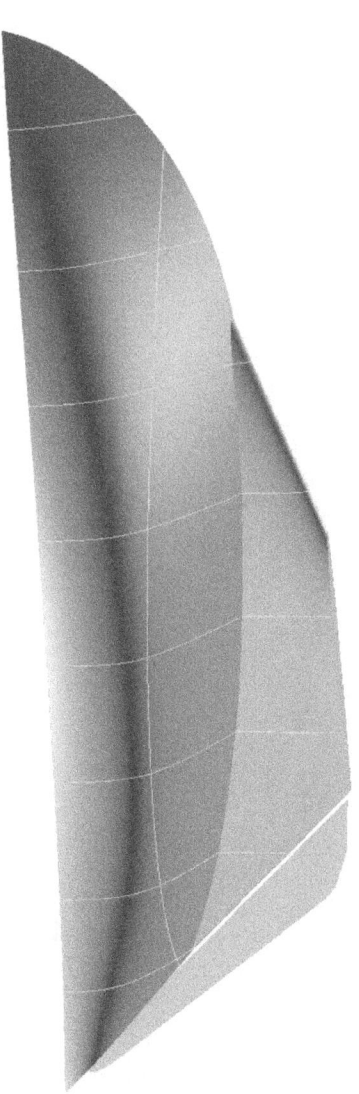

Two Surface Bezier Design:

In a Double surface design, the upper compound surface runs from the sheer-line around the turn of the bilge to a theoretical chine line, flowing tangent and seamlessly into a Developable or Elementary surface to the center line of the hull.

The theoretical Chine Line that separates the two surfaces is a visible demarcation in the hulls 'Lines' drawings, however, in the finished hull it is invisible too the eye.

Shown in the below drawing are the Transverse Frames, Design Waterline, and the Theoretical chine line.

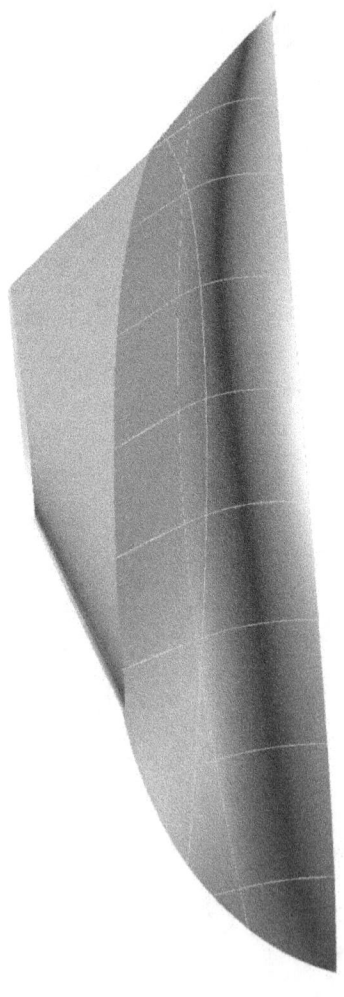

Preference for 'Single Surface' Hull Configurations:

Since building the 'Double Surfaced' Bezier 12.5, I have come to the realization that combining a True Round surface with a Developable bottom surface serves no real purpose, however the option is still available if one so desire's.

I now feel, that a 'Single' surface True Round design configuration has many advantages over the 'Double' surface hull configurations that I was compelled to redesigned the 'Double' surface prototype version of the 'Bezier 12.5' to a 'Single' surface configuration.

The advantages of a 'Single' Surface Design over a 'Double' Surface design are:

* The Design Process itself was significantly simplified with the absence of a Theoretical chine.
* The longitudinal framing system is less problematic.
* Welding is simplified - No Longitudinal shell plate weld seams.
* One plating method.

In the redesign of the 'Bezier 12.5 from a 'Double' to a 'Single' hull configuration several hull features were strictly retained.

* Shape and position of the Sheer-Line.
* Shape and position of the Fairbody Line.
* Location and shape of the Bow.
* Position, shape, and angle of the transom.
* Location and configuration of the Keel and Rudder.

Keeping the above geometry of the hull resulted in little to no change in displacement, Centers of Buoyance, Centers of Gravity, Stability, and Design Coefficients.

The Below two Illustrations represent the 'Master Curves' that define both versions of the Bezier 12.5 overlayed onto each other. It can be seen that there is little difference in hull shape between the two hull configurations.

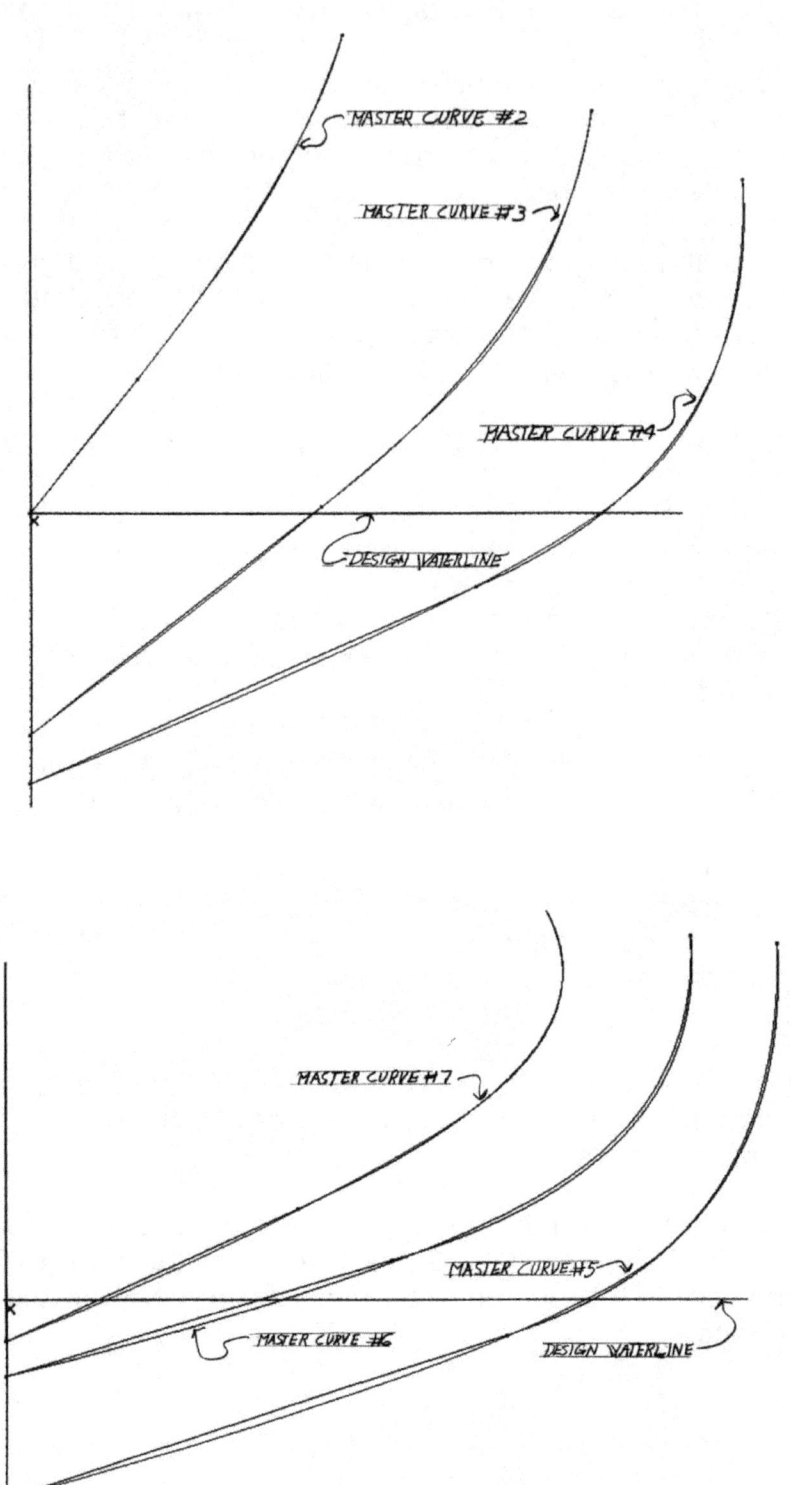

MASTER CURVE #2

MASTER CURVE #3

MASTER CURVE #4

DESIGN WATERLINE

MASTER CURVE #7

MASTER CURVE #5

MASTER CURVE #6

DESIGN WATERLINE

Three Surface Bezier Design:

Triple Surface Bezier Designs place a True Round surface between Developable or Elementary surfaces. The middle surface can be either:

- **Free-Formed**
- **Single Constant Radius – Also know as a 'Radius Chine' hull.**

The Free-formed Version:

The Bezier 28 will be use to illustrate the concept of a Triple surface Bezier design.

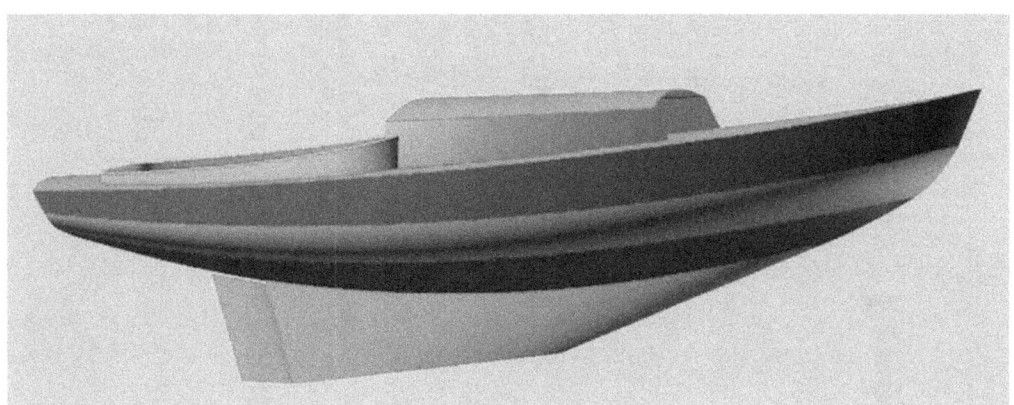

To demonstrate why the center True Round surface of the 'Bezier 28' is indeed a True Round or Warped surface, not just a 'Single Constant Radius' inherent to a 'Radius Chine Design'. I have estimated a radius that would approximately the free-formed curves at transverse frames: Three, Seven, and Eleven for the Free-formed Bezier 28.
In the, below Illustrations, Frame three would have a radius of 27", Frame seven would ha a radius of 18", while Frame Eleven would have a radius of 13".
These varying radii, illustrated below, at the aforementioned transverse frame in girth at different intervals along the length of the hull indicate that the hull is indeed a True Round surface.

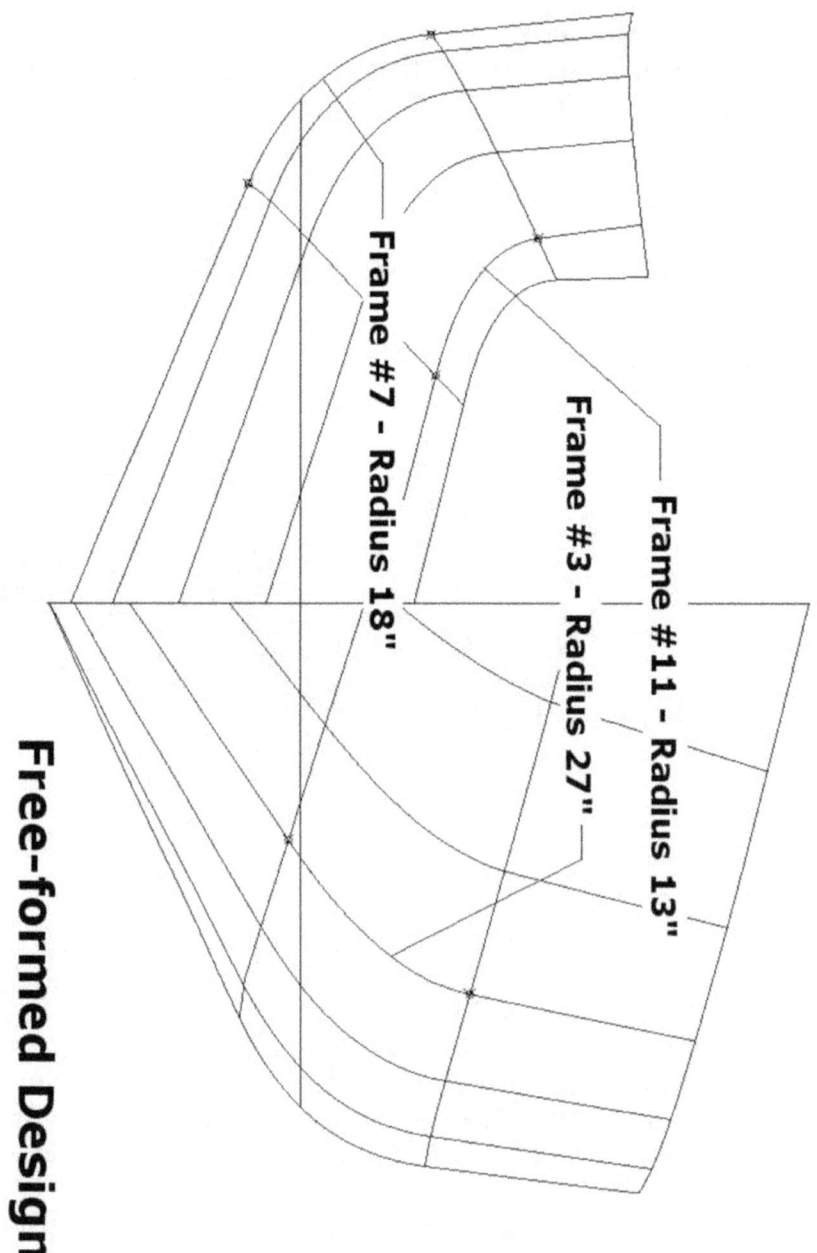

Free-formed Design

Frame #7 - Radius 18"

Frame #3 - Radius 27"

Frame #11 - Radius 13"

The Radius Chine Version:

It is commonly thought that a 'Radius Chine' hull is not really a 'True Round Hull'. That a 'Radius Chine' hull only imitates a 'True Round hull'. It is my contention that a 'Radius Chine' hull is a 'True Round' hull. Here is why!

When a 'Radius Chine' hull is on the Designer's drawing board the hull surface between the developable surfaces is intentionally designed to be a 'Single Constant Radius' in Girth which, as when seen, is one direction of roll.
The other direction of Roll, like the Free-Formed version, curves along the length of the hull, therefore, if a surface curves in two directions by definition it is a 'Warped' or 'True Round' surface.

'Radius Chine' designed hulls actually have an advantage over 'Free-formed' designed surfaces in that the shell plate patterns are **'Rolled Formed'.**

On the negative side, **'Radius Chined'** designs are subject to hull form design restriction due to the design process itself.

The following three (3) **'Radius Chine'** designs are based on the 'Free-formed' version of the Bezier 28 hull design.
- The 8" radius is about the lowest radius a 'Designer' would choose.
- A 16" radius is about the largest that could be used, since the largest radius is governed by the width of the transom.
- The 12" radius is used to reference the middle ground between the upper and lower limits.

While I could comment extensively on the 'Three' surface designs, I leave it to the Reader to evaluate all the 'Three' surface designs and come to their own conclusions on which is best?

**Free-formed
or
Radius Chine**

The 8" Radius Chine Design

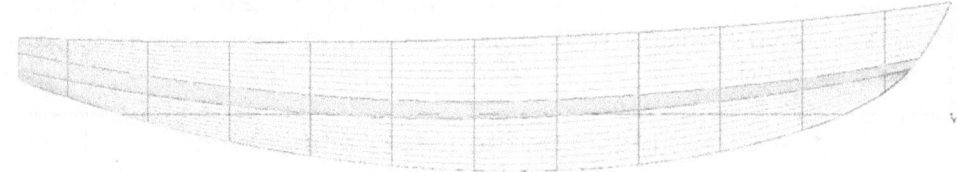

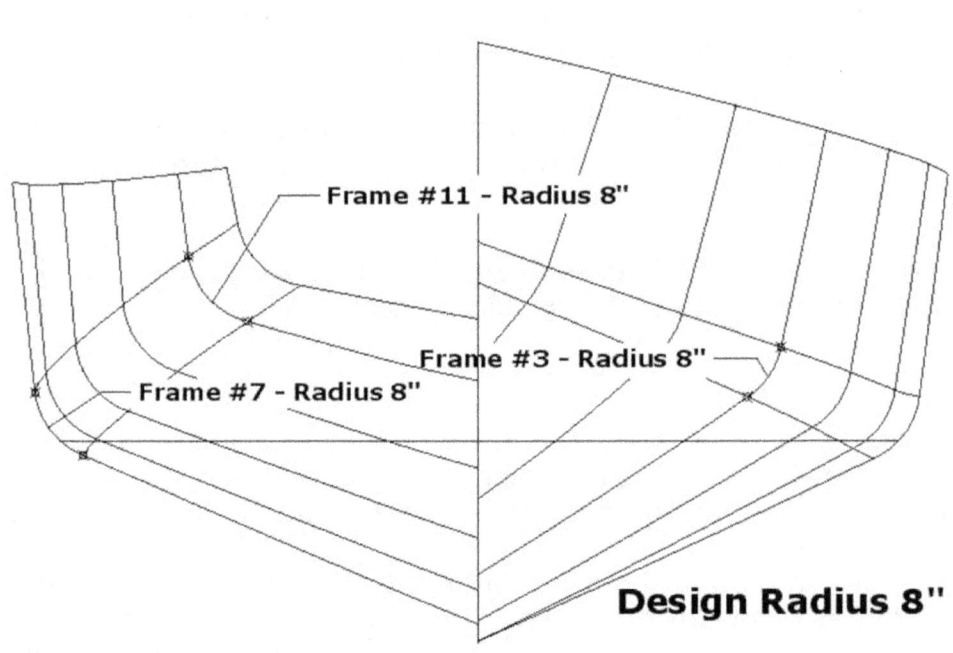

Frame #11 - Radius 8"

Frame #3 - Radius 8"

Frame #7 - Radius 8"

Design Radius 8"

The 12" Radius Chine Design

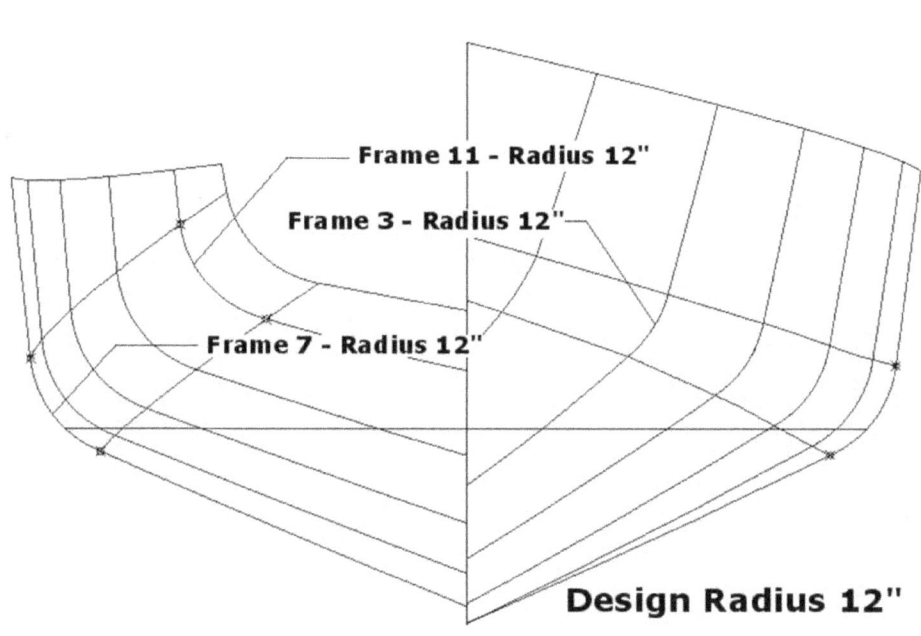

Frame 11 - Radius 12"

Frame 3 - Radius 12"

Frame 7 - Radius 12"

Design Radius 12"

The 16" Radius Chine Design

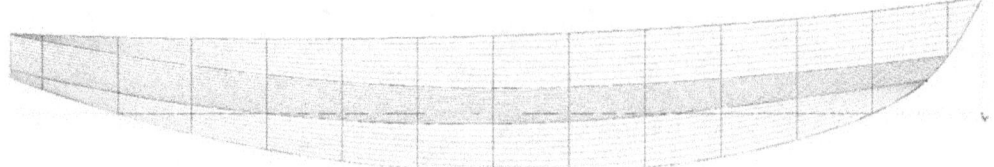

Notes

Bezier 12.5 Photo's

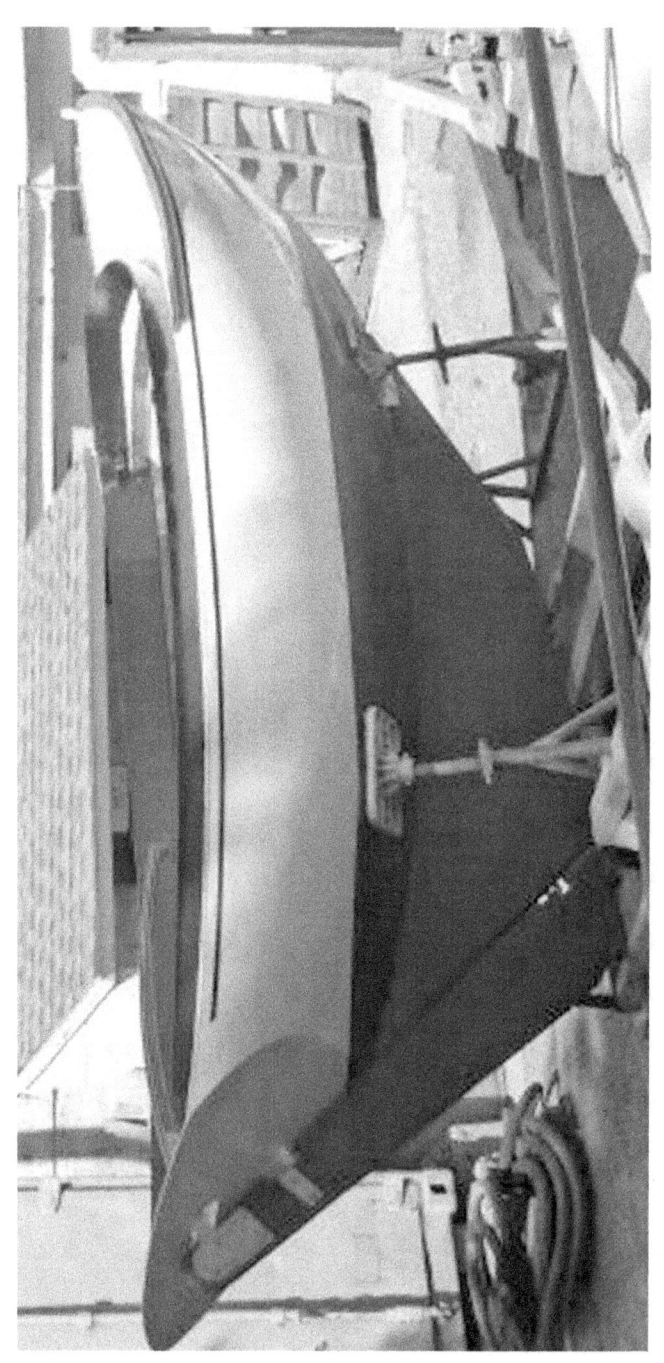

Notes

Other Bezier Designs

Bezier 28

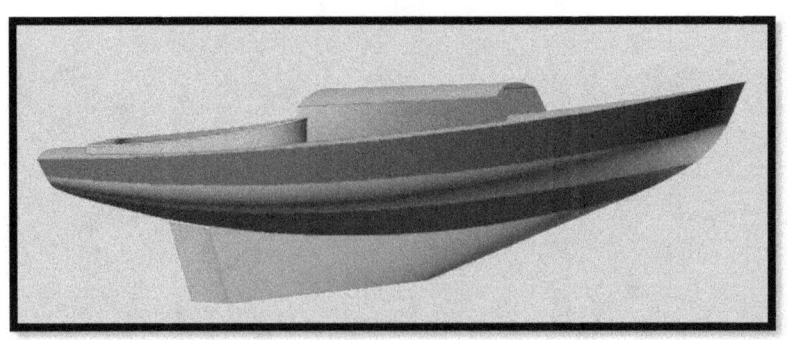

Bezier 35

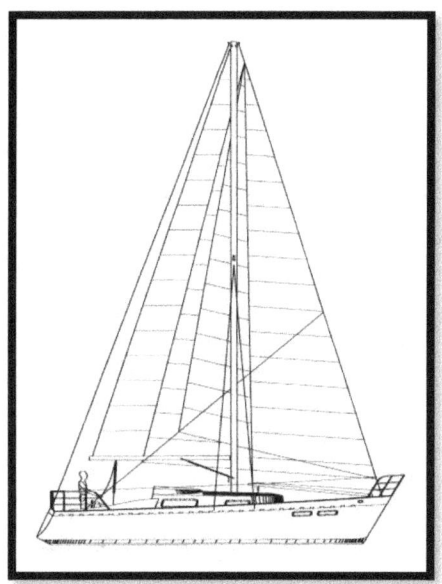

Bezier 12.5

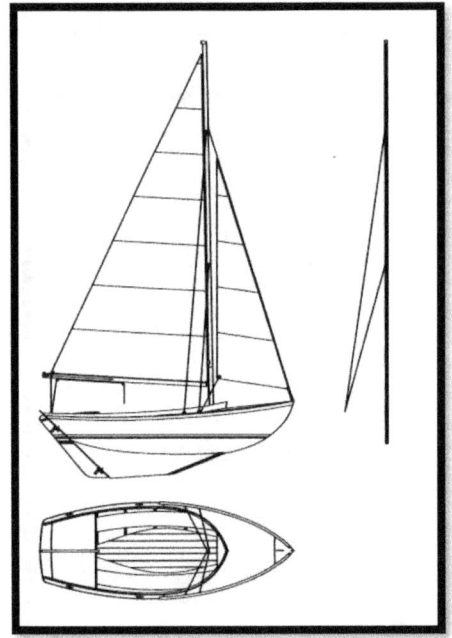

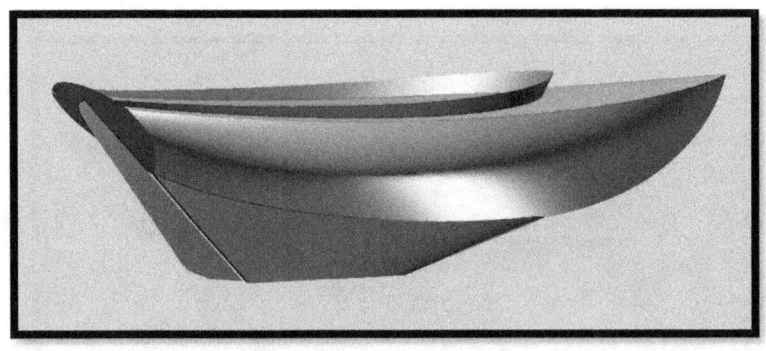

Bezier 34

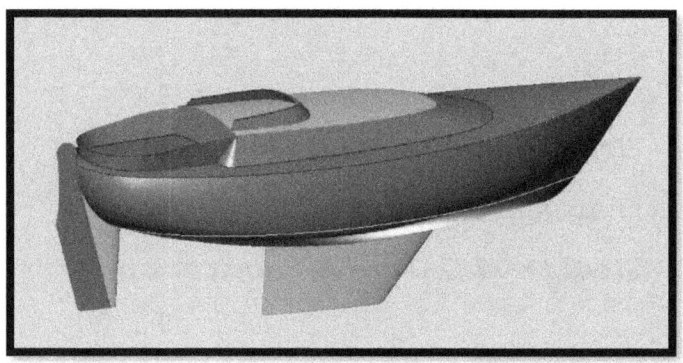

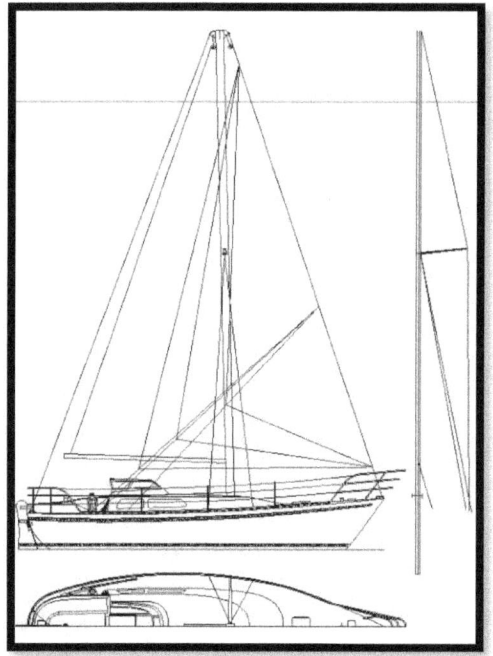

Other Books

(Available at Amazon)

True Round Metal Boat Design

Fabricating the Hull Integrals

Steel Mast Design & Fabrication

www.ingramcontent.com/pod-product-compliance
Lightning Source LLC
Chambersburg PA
CBHW080658190526
45169CB00006B/2173